LEIBNIZ AND DYNAMICS

PIERRE COSTABEL

Leibniz and Dynamics

THE TEXTS OF 1692

Translated by
Dr R. E. W. Maddison, F.S.A.
Librarian of the Royal Astronomical Society

HERMANN: PARIS
METHUEN: LONDON
CORNELL UNIVERSITY PRESS: ITHACA, NEW YORK

Translated from the original French text *Leibniz et la dynamique*,
published by Hermann, Paris, in 1960, in their series *Histoire de la Pensée*

English translation © 1973 by Hermann, Paris
Published in France by Hermann
293 rue Lecourbe 75015 Paris
Published in Great Britain by Methuen & Co Ltd
11 New Fetter Lane, London EC4
Published in the USA, Canada and Mexico by Cornell University Press
124 Roberts Place, Ithaca, New York 14850

ISBN (Hermann) 2 7056 5673 1
ISBN (Methuen) 0 416 77000 2
ISBN (Cornell) 0–8014–0775–3

Library of Congress Catalog Card Number 72–13060

Foreword

The present study would never have seen the light of day had it not been for the encouragement and advice of the late M. Alexandre Koyré, director of studies at the *École Pratique des Hautes Études de la Sorbonne*. His lectures on the history of science have enabled us to make a careful study of the fundamental questions discussed during the course on the seventeenth century and to experience the benefit of a collective research worthy of higher education.

By reason of our own individual work in the course, our study was first of all directed to a more technical subject, namely, the problem of impact as investigated by Huygens and Mariotte. It was during our fruitless search for documents relating to Mariotte's experiments that we came across two Leibnizian manuscripts in the archives of the *Académie*, and that discovery caused us to change our plans. We have no regrets for having done so.

The manuscripts in question are two copies, dating from the year 1692, of texts dealing with mechanics. Examination of internal and external evidence, identification of the copyist, investigation of the circumstances of, and the motives for, making the copies led us to a careful study of an imperfectly known story, namely, the difficulties encountered by Leibniz in the French intellectual circle. A patient, detailed study enabled us to discover elucidation of the more general and extended theses on the internal coherence of the thought of Leibniz. The result confirms a general view, to which many minds are nowadays tempted to limit the benefit of detailed study. It has been our one ambition to show through our work that such a hasty conclusion is not justified. Grand syntheses and general ideas are precious and indispensable. Nevertheless, they have no firm substance without profound and lively knowledge which is acquired patiently by contact with the great men of the past. The tact, mistakes and inadequacies of Leibniz make him most human and bring him close to us; the depth of his analysis and his tenacious application in following to the end

7

the lines traced by metaphysical principles make him even greater, and force us to realize the difficulties he had to overcome, which difficulties were related not only to his period and to his environment, but which are perhaps eternal.

Everything that restores life and saves us from eternal abstract discussion of ideas is of great value. If we have been able to provide some slight proof thereof, then our thanks on that account are due to the lectures of M. Alexandre Koyré.

We have to thank M. Joseph Ehrenfield Hoffmann who saved us much time in seeking the first manuscript by providing an immediate reference. Special thanks are due to M. André Robinet for information kindly placed at our disposal. The appearance of M. Robinet's thesis on Malebranche and Leibniz was a very great help when we were finally assembling the necessary facts for the account in our first chapter, which aims at being exhaustive.

PIERRE COSTABEL

Contents

Introduction

Much has been written about Leibniz as indeed about other important philosophers of the seventeenth century; and it will be readily understood that an individual whose initial training was in mathematics makes no claim to add to the learned commentaries from which he himself derived knowledge the complexity of which he was quite unaware some years earlier.

If he indulges in these personal reflections, it is because his own experience would seem to have some value for those of the same intellectual outlook. The modern scientific mind is in general very satisfied with the independence of knowledge with respect to all philosophy, each of the disciplines constituting this knowledge being gradually constructed as it were homogeneously with the purpose of rejecting every external element. It considers itself able to judge the works of the past by reference to the present state of facts. When it approaches history, it does so, for the most part, in order to discover therein those emergences which it justifiably calls positive heritage, and if objectivity leads it to note the metaphysical grounds which were the prime movers of research among the great forerunners of the seventeenth century, then they are rarely accorded more than polite attention. To reduce these grounds to the function of catalysts is a natural tendency for a mind that no longer belongs to the same world as these great forerunners; and if this be a very regrettable state of affairs, there is no need to be shocked; it is far better to try and restore, as far as possible, an understanding of the vanished world.

As M. Guéroult has said,[1] 'One has only to refer to a list of the early works of Leibniz to be convinced of the varied sources of the subjects (logic, mathematics, physics, ethics, religion, theology, philosophy, etc.), and on analysis to discover all of them in each, though from different viewpoints. The themes develop together, re-

1. M. Guéroult, *Dynamique et Métaphysique Leibnizienne*, Paris, Vrin, 1934, p. 2.

11

acting simultaneously on each other, so that as regards his philosophical conceptions it could be said that Leibniz asserted everything in the universe to be related and to agree. . . .' That is true of Leibniz to a very special degree, and is true also of those, such as Descartes, Pascal, Newton, etc., who are regarded by science as the great precursors. To note that they were both philosophers and *savants*, and to note that they found powerful motive forces for their scientific research in their philosophy merely leaves us on the threshold of a genuine and profitable understanding of their works. If, from their point of view, though expressed in different ways, everything is related and everything fits together, then doubtless they were aware of fundamental problems—eternal problems—the existence of which we have perhaps forgotten.

Mathematically trained minds are greatly surprised when, for example, the difficulties of the concept of a differential are pointed out to them. We remember what was said to a learned mathematician to whom these difficulties had been put: 'If the student does not understand how it is possible to reason and calculate with quantities that are essentially variable and indeterminate, and which are neither nothing, nor something, then he must have faith. Be confident, follow the logic of the rules you have been given, calculate, and then you will find no contradiction.' If the mathematicians have lost all hope of making the absence of contradiction itself a subject of proof, they continue, nevertheless, to find a criterion of truth in it—and rightly so; but they believe too easily that former difficulties are thereby purely and simply eliminated. Leibniz was one of the first to devote himself to the calculus, and he was able to say with all due modesty with respect to the differential calculus that it was 'his' calculus.[1] Nevertheless, he never lost sight of the contradictions introduced into natural science by the mathematical postulate of infinitely divisible continuity.

The difficulties are obscured, or seem to vanish, when the field of vision is limited. However, nothing is more dangerous, for the very movement of existing thought hardly admits of any limits. The philosophical *savant*, who in the seventeenth century knowingly kept all his researches within the vast framework of universal knowledge is certainly preferable to the strict and specialized scientist of the twentieth century, who moves easily in his own particular field and

1. See especially Gerhardt, *Leibn. math. Schr.*, vol. VI, p. 507 and *infra* p. 84, note 1.

believes that he can become sufficiently competent in the others simply by enlarging his range of interest. There is no doubt but that he has the advantage of not passing over any difficulty in silence, and he invites us to learn with appropriate seriousness.

The science of motion holds a position of prime importance in the philosophy of Leibniz. This general statement is not enough as such to characterize the reason for Leibniz's position. Since the time of Aristotle every attempt at a rational construction of a knowledge of nature has hit upon motion as an essential feature, and Leibniz entered a world profoundly apprised of what is called mechanism. He did not invent the principle according to which everything in nature must be explained '*per magnitudinem, figuram et motum*'; nor did he invent the principle according to which motion is the major element that 'engenders in merely extended, strictly homogeneous and indeterminate primary matter those determining factors which decide the size and shape of bodies'. It was a Cartesian idea to deduce the properties of matter from the abstract properties of our notion of 'extension' (continuity, perfect homogeneity, absence of limit), and to try and find in motion the principle of division into innumerable parts which gives rise to real bodies in matter. It was also a Cartesian idea to derive all the so-called secondary, tangible properties of bodies, such as light, colour, heat, gravity, etc., from motion.

In his youth, Leibniz was subjected to the influence of Cartesian doctrine much more than he was willing to admit to himself. It is true that he read Descartes at a late date.[1] About 1669, when he was putting his thoughts together for the '*hypothesis physica nova*', he knew Descartes only through secondary works. Nevertheless, he immediately found himself in complete agreement with him as regards the same care for rigorous elaboration of a geometric kind for universal knowledge of nature. To start from perceptible phenomena, real and apparent, in order to explain 'true' laws, as is done by many 'moderns', did not seem to him to be a good, sound method, even when the argument agrees with the observation. The good method is that for which philosophers have acquired a liking from the *Elements of Euclid*, and though it be difficult, Leibniz wished to put it into practice.[2] Broadly speaking, he is essentially a *rationalist*.

1. A. Hannequin, *La première philosophie de Leibniz*, p. 22 (Letter to Fabri). Gerhardt, *Die phil. Schr. v. Leibniz*, vol. IV, p. 247; see *infra*, p. 45.
2. *Meditationes de cognitione, veritate et ideis*, 1684, in Gerhardt: *Phil. Schr. v. Leibniz*, vol. IV, p. 426; *Theoria motus abstracti* (*ibid.*, pp. 234, 239–240).

At the same time he is a sincere believer. Though the human mind can conceive and imagine things to be possible, it is powerless to call them into existence; God is necessary for a universe that does not have existence without creative and immediately efficient thought. If the creative movement definitely proceeds from God, then the action of the divinity continues and extends itself as it were of its own accord; and in order to explain the universe there is no need to introduce an extraordinary, constantly renewed assistance on the part of God. We should greatly deceive ourselves by unduly magnifying the part played by the religious element in the genesis of Leibnizian doctrine. No doubt, as Paul Mouy has said,[1] 'his profound intent, which he realized better and better, was to establish a conciliation that would save all the values, not only those of science but also those of law and theology as well, and to derive from mechanism itself the means of proving the existence of the soul and of God, to make God confessed through matter in order to confound the atheists', and it is not an insignificant paradox, that the Protestant Leibniz is endowed with a Catholic optimism as regards the positive value of the human faculties. The intelligibleness they allow us to attain is not a misleading mirage; the human mind is not the divine power, but it is not affected by a vital infirmity that vitiates beforehand all its various proceedings. It possesses a real adequacy for the pursuit of knowledge, and the God of Leibniz, as well as of Descartes, is a geometer.

It would be pointless to seek a distinction between these two great minds by contrasting two kinds of rationalism. On the contrary, they are closely united by their fundamental position.

All the same, there are some differences. The comments which accompany the writings of Descartes on mechanics are evidence of his awareness of the necessary difference between abstract laws deduced from logical speculation and the laws of observable phenomena. For example, the laws of impact were established in an imaginary world of perfectly hard bodies and in a vacuum. Such conditions are unreal. The laws of the inclined plane, the pulley, and the lever assume complete absence of resistance to motion. In fact, the roughness of the surfaces, the stiffness of the cords, etc., introduce friction which upsets the theoretical deductions.[2] There are, certainly, some incon-

---

1. Mouy, *Le développement de la Physique Cartésienne, 1646–1742*, Paris, Vrin, 1934, p. 218.
2. Letter to Mersenne, 27 May 1638. Letter to Constantijn Huygens, 5 October 1637, *ed.* Adam Tannery, vol. II, pp. 142, 432–443.

sistencies for certain theoretical laws which are difficult to explain simply by disturbances brought about by phenomena neglected as a first approximation. Such is the case with regard to the famous fourth 'Rule' which states that a stationary large body cannot be set in motion by any means by a smaller body colliding with it. It seems that more would be needed to move Descartes. However, Descartes asserts that with time and patience, of which he is often in need, he would solve the difficulty and would find a rational explanation of all the established discrepancies.[1] His boldness in firmly maintaining, even against seeming unlikelihood, the pre-eminence of rational knowledge over concrete physics depends on an act of calm and serene faith. The attitude of Leibniz was different. The strange conflict that is occasionally revealed between the laws of abstract motion and the laws derived from experiment gave him no rest. The latter must be derivable from the former; a rational picture of the way in which God found a way of establishing harmony between geometric exactitude and physical reality must be found.[2] The rationalism of Leibniz is not satisfied by finding loopholes; it is eager to go further and deeper.

We said above that there were some slight differences. They are, evidently, of undeniable practical importance. Without them, we should not be able to understand what it was that separated Leibniz from his great predecessor.

In 1693, in his famous letter to Nicaise, Leibniz wrote: 'I am sure that if M. Descartes had lived longer, he would have given us an infinite number of important things.[3] Which shows, either that it was his genius rather than his method which caused him to make discoveries, or that he has not published his method. Indeed, I remember having read in one of his letters that he merely wanted to write a dissertation on his method and give some examples of it, but that it was not his intention to publish it. Consequently, the Cartesians who believe that they are in possession of their master's method are greatly deceived. However, I imagine that the method in question was not so perfect as we would believe it to be. I judge him by his Geometry. That was undoubtedly his strong point; nevertheless we know

1. Descartes, *Principes de la Philosophie*, Amsterdam, Elzevier, 1644, Part 2, art. 53 and 64; Part 4, art. 188.
2. *Theoria motus abstracti*, Gerhardt, *Die phil. Schr. v. Leibniz*, vol. IV, p. 237.
3. *Journal des Sçavans*, 13 April 1693.

now that it was infinitely lacking in that *it does not go so far as it should and that he said that it did*. The most important problems require a new analytical method quite different from his, of which method I have myself given some examples. . . .'

At the time corresponding to the texts which are given in this present study Leibniz was aware of having surpassed Descartes, of having discovered through analysis and the differential calculus a secret which had escaped the master, and naturally, what is more, his followers, too. This secret is that the purpose of mathematics consists essentially in speculation on ratios and proportions and in a study of relationships, so that the realm of the infinitely small, notwithstanding logical difficulties, ceases to be unintelligible and can, on the contrary, be subjected to rational organization. If Leibniz discovered the secret in 1675 during his stay at Paris, when Huygens, his instructor in mathematics and mechanics, allowed him to fill the gaps in a superficial training by means of appropriate reading, and made a mathematician of him,[1] then it was because the ground had been particularly well prepared by previous philosophical reflection. Consequently, it is impossible to understand why and how Leibniz parted company with Descartes, why and how his rationalism goes beyond Cartesian rationalism in its true outline, without referring to the works that immediately preceded the acquisition of a real knowledge of mathematics by the German philosopher.

The *Theoria motus abstracti* was the first of two works written by Leibniz in 1670 and put forth under the title of *Hypothesis physica nova*.[2] It is the basic treatise of the vast universal system with which the young philosopher never ceased to be concerned. The work opens with the '*fundamenta praedemonstrabilia*' inspired by Cavalieri's geometry of indivisibles, the purpose of which is to explain the nature of the Continuum. To imagine, as did Gassendi in order to explain the various degrees of motion (acceleration), that longer or shorter intervals of rest enter into the nature of motion, merely shifts the ground of the problem. The 'grains' of motion, however small they may be, are still motion, and the original problem emerges at infinity, if we confine ourselves to a discontinuous structure.

Motion is, therefore, something continuous, and the nature of

---

1. Huygens, *Œuvres complètes*, vol. VII, p. 244, note 12.
2. Gerhardt, *Die phil. Schr. v. Leibniz*, vol. IV, pp. 223–240; see Hannequin, *La première philosophie de Leibniz*, pp. 60–61.

every continuum, as Cavalieri said, is to be divisible and infinitely divisible. No continuum exists that does not have an infinite number of parts, and yet none of these parts can be considered as indivisible without our being inconsistent. If every portion of space, time and motion is still space, time and motion, then each portion must still be indefinitely divisible. There is no minimum extension. However, what is to be said regarding the 'beginning' of a body, of duration, of motion? Such a beginning belongs to space, time and motion without being itself divisible, for the idea of a divisible beginning is contradictory. Consequently, there are many indivisibles constituting space, time and motion; nevertheless they are heterogeneous with respect to that which they constitute seeing that extension cannot be 'assigned' to them without our falling into one contradiction from another. That has been known since the time of Zeno of Elea,[1] but if we wish to extricate ourselves from the difficulties of such an old problem, we must reconcile ourselves to reasoning on those strange things, the point, the instant, the *conatus*, whose nature is to be 'unassignable'.

Leibniz had no hesitation in doing so. He borrowed both the term and definition of *conatus* from Hobbes: '*Conatum esse motum per spatium et tempus minus quam quod datur....*' The rudiment of motion therefore associates under the term *conatus* an *unassignable* or infinitely small duration with a space of the same kind. The concept is completed, as Hannequin has rightly pointed out, by a consideration of velocity. 'In the same way that there are uniform motions of different extent, or what comes to the same thing, velocities of different degree, so *conatus* can exist in forms that differ amongst themselves as do velocities.' In order to make a comparison between various *conatus*, it is a fact that we must assume uniform passage of time itself, and Leibniz readily allows it, even though he is aware of the logical difficulty thereby raised. However, this hypothesis having been laid down, the concept of *conatus* acquires a rather clear meaning. It is the infinitely small element of motion that finds its determination in the velocity of a uniform, rectilinear motion coinciding with the same space and with the same infinitely small duration.[2]

We have already noted the influence of Hobbes on the thought of Leibniz; and it must be considered further in order to understand the

1. See Pierre Costabel, *Le Mouvement*, Paris, Encyclopédie Clartés, 1956, vol. XVI.
2. Hannequin, *La première philosophie de Leibniz*, pp. 80–84.

plan of the *Theoria motus abstracti*. '*Causa motus nulla esse potest in corpore, nisi contiguo et moto*', was the principle of Hobbes, which excludes from the science of motion all action at a distance.[1] Leibniz derived inspiration therefrom all the more readily seeing that he did not accept the vacuum, and like Descartes was in favour of a physics of the plenum, but of a plenum which can be said to be full of motion. Henceforth, the two essential problems to be resolved were those of the compounding and communication of motions.

'By the kinematic compounding of two or more motions into one only,' said Hannequin, 'one thing seems at first sight to distinguish the impact of bodies: it is, that in the latter case the modification of the combining motions takes place suddenly and as it were in an instant, whereas in the former case the moving body is regarded as the point in which two or more motions unite during a finite time, as though it did not cease to be animated simultaneously by different motions. In both cases, we are concerned with motions that combine and it is a matter of finding out how they are compounded.'

The *Theoria motus abstracti* is, therefore, definitely a treatise on the compounding of *conatus*, and the laws of impact, which are of prime importance for a doctrine in which that is the sole mode of exchange of motion, are reduced to algebraic summations of *conatus*. The mass of the moving bodies is not involved at all. We should not be surprised thereat. In the absence of motion a body is not distinguishable from the place it occupies in space; without velocity a body is nothing. It is perfectly logical for the size of bodies to be of no importance, and for all power in this universe to be expressed in the *conatus* and to be measured by velocity.

If the doctrine be certainly coherent, it has, nevertheless a very unfortunate consequence. By the laws of compounding it follows that in all encounters there is a removal of velocity. Considering the universe as a whole, it seems, therefore, that motion must destroy itself and the universe return to emptiness. Leibniz was aware of this unescapable conclusion[2] and ended his first attempt at a rational construction with a feeling of fundamental uneasiness.

The system of Descartes did not meet with such misfortune. In creating the universe, God put into it the same amount of motion as of

1. Hannequin, *ibid*, p. 87.
2. *Phoranomus*, 1689, in Gerhardt, *Archiv der Gesch. der Phil.*, vol. I, pp. 575 *et seq*; Gerhardt, *Die phil. Sch. v. Leibniz*, vol. VII, p. 260.

rest, and the amount of motion is preserved in such a way that the universe continues its existence without requiring the constant and regenerative assistance of the divinity. Consequently, it is essential to establish a law of universal conservation in any attempt at a rational explanation of the universe. That is the lesson that Leibniz was to put beyond doubt. When the long chain of logical deductions starting from some simple principles has been unfolded, the presence or absence of a universal conservation constitutes a crucial criterion for the truth or error of the whole system.

When Leibniz arrived in Paris in 1672, he had not yet acquired so clear a vision of things. He was eager to improve his mind and to put to the test the soundness of some fundamental points in his first analysis by contacting some of the great masters. He was not disposed to abandon the demands of his reason. In contrast to Cartesianism, for which anything that is clearly understood is intelligible and which gives in short the path to intuition, Leibnizian rationalism requires an exhaustive analysis of the existing conditions of the concepts employed. In this perspective, the meeting with Malebranche ended in a setback. It was a question of knowing if space were distinct from matter. So far, Leibniz had adopted the Cartesian position on this point: matter is identical with extension. However, he wanted a proof. Malebranche merely outlined one. An empty space is inconceivable. The concept of space implies distinction between the constituent parts; therefore these parts are separable; therefore they are mobile; therefore they are material. Leibniz was not satisfied. 'It remains for us to prove', he said to Malebranche, 'that two things, such as the parts of space are, have no requisites. For me, every thing that can be produced has requisites outside of itself, namely those that conspired to produce it. Now, parts of space are produced by the motion of the body that cuts across it; therefore the parts have requisites'.[1] Consequently, Leibniz was obliged not to admit that the parts of space are absolute things. The concept of them is not simple, and the confrontation which Leibniz looked forward to on the philosophical plane ended by leaving him rather worried.

It was quite different on the mathematical plane. Discovery and enlightenment resulted from the meeting with Huygens and the reading carried out under his advice.

Leibniz discovered the principles of infinitesimal analysis.

1. See Letters I, II, III in Gerhardt, *Die phil. Schr. v. Leibniz*, vol. I, pp. 315–361.

Cavalieri's geometry by which he was inspired had the serious inconvenience of necessitating a judicious compromise in practice in the solution of problems. From then onwards, Leibniz knew that it was not necessary to devise indivisibles, even those having no extension, in order to subject the realm of the infinitely small to calculation. The study of ratios and proportions, which constitutes essentially the science of relationships in mathematics, is applicable just as well to the realm of the infinitesimal as to the realm of the finite. By taking some precautions, a rational means, which has nothing in common with the inaccurate methods of Cavalieri, was then found in order to explain the genesis of realities. The intelligibility of this genesis no longer involves embarrassment in the difficulties of an indefinitely divisible continuum and the existence of indivisibles having no extension. Leibniz, having become a mathematician, was able to retain the word *conatus*; he no longer had need in mathematical usage of the notion that gave it substance or real existence. The only requirement for that usage is the ratio between distance and time, both being infinitesimal, that is to say, the velocity of the uniform motion equivalent to the motion considered over an infinitely small duration of time. The enlightenment of Leibniz was to the effect that it was not necessary definitely to have taken sides on difficult questions in order to calculate and to calculate accurately. Mathematics with infinitesimal analysis is an instrument of choice.[1]

On the other hand, as Paul Mouy says, Huygens, 'if he be a Cartesian even with its inadequacies, nevertheless brings the mentality of a calculator to Cartesian physics. His work starts from Cartesian principles, but he aims at formulating a general law which expresses itself through an algebraic construction, without typical hypotheses and without mechanical models. His aim is to calculate, and if his calculation can be expressed uniquely by one construction, and not by algebraic formulae, it is the result of his technique, and not the result of a turn of his mind.'[2] This turn of mind perfectly suited the pupil that Huygens met in Leibniz, a pupil who so quickly surpassed his master in the field of mathematical technique. We can understand why Leibniz was able at a later date to give Huygens the honour of having 'been the first to purge the doctrine of motion from

1. See Brunschwigg, *Étapes de la philosophie mathématique*, Paris, Alcan, 1912, Section B: the geometry of indivisibles and the Leibnizian algorithm.
2. See Mouy, *Le développement de la Physique Cartésienne, 1646–1742*, Paris, Vrin, 1934, p. 200.

all fallacious argument'. Thanks to the intervention of Huygens, Leibniz not only discovered the secret of the theoretical power of mathematics, but in addition he was accustomed to consider its application to the science of motion.

His stay at Paris, his relations with Huygens and Malebranche, his discovery of the infinitesimal calculus were decisive events for the thought of Leibniz. All the same, we should be wrong in believing that after the setback of his philosophic encounter with Malebranche Leibniz resolutely set himself in a strictly and purely mathematical perspective. The philosopher abandoned none of his rights.

No doubt, and this is very important, he learned through contact with Huygens to give suitable attention to the laws 'that experiment teaches', to the concepts of mass and of elasticity which alone are capable of putting in order and accounting for observable realities. It was not enough for him to be, henceforth, in possession of a more suitable mental equipment, which he was moreover engaged in perfecting for his mathematical renderings. It was furthermore necessary for all that to be combined with the demands of metaphysics.

In the *Phoranomus*,[1] written at Rome in 1689, Leibniz has himself explained the course of his reflections: 'When I was aware only of the *juridictio imaginationis* with regard to material things, I thought that it was not at all possible to assume natural inertia in bodies. . . . Recognizing only extension and impenetrability in matter, in other words only *impletio spatii*; understanding nothing more in motion than *mutatio spatii*, I considered that the difference at every moment between a stationary body and a moving body to consist in this, that the moving body possesses a certain *conatus* or tendency to start its course . . . I did not see how a *conatus* could be destroyed in nature or removed from a body.' No doubt, he realized that a contrary *conatus* would have a balancing effect and would decrease the motion; but for the laws of the communication of motions he was unable to deduce from the simple geometric concept of a body 'any reason for excluding or simply limiting the *conatus* in the body which ought to receive it'. Whence we have a purely kinematic science, making no call upon the size and mass of bodies, and manifestly conflicting with the results of practical science. Whence the impossibility of accounting for resistance to motion which is the fundamental residuum of the

---

1. *Phoranomus*, Gerhardt, *Archiv der Gesch. der Phil.*, vol. I, pp. 572–580.

most ordinary, rough experiment. Whence the futility of assuming a directing principle for future mechanical actions in a universe reduced to mathematical concepts, and of assuming in that principle an equilibrium between losses and gains of motion. 'For these and many other reasons', said Leibniz, 'I concluded that the nature of matter was not yet sufficiently well-known to us, and that we could not account for the force of bodies if we did not put in them something other than extension and impenetrability. The mechanical principles and laws of motion sprang, in my view, undoubtedly from the necessity of matter, but in any case from a superior principle independent of intuition and mathematics.' The last word in this discussion in which Leibniz was occupied through his own reflections was that 'in order to escape from the labyrinth', there is no 'other Ariadne's thread but the *evaluation* of forces, subject to the supposition of the *metaphysical principle* that the total effect is always equal to the entire cause'. Salvation is possible only through informed mathematical analysis, directed and enriched in its purpose by metaphysics.

There is no mathematical analysis without co-ordination of concepts and symbols. Leibniz found himself confronted by the mass-velocity method of co-ordination represented by a product called quantity of motion. Strict followers of Cartesianism were not the only ones to make use of it. In a physics that takes no account of the infinitely small and which recognizes only finite variations, the instantaneous passage from rest to motion, or from one degree of motion to another, seems naturally to require characterization by a double proportionality between the quantity of matter and the acquired velocity. What, then, is more natural than the equivalence of $(m, 2v)$ and $(2m, v)$, which will be interpreted by saying that if twice the force is required to give the velocity v to a body of mass $2m$, then the same force is sufficient to give a body of mass m the velocity $2v$. It is true, that Descartes, as we shall have occasion to see more fully later, was more cautious and, in particular, did not subscribe to this simultaneous doubling of force and velocity; but, in bequeathing to posterity a concept of velocity deprived of the infinitely small coupled with the principle of the universal conservation of the quantity of motion, he has the responsibility for identifying the product mv with the measure of the dynamical principle or of the force.

For Leibniz, on the other hand, the idea of an instantaneous transmission of motion was absolutely excluded. There is always production or progressive and continuous communication, even though they may be more or less rapid, and a finite variation in velocity requires a finite time during which the matter manifests resistance to the phenomenon of motion. The concept of mass is necessary to ensure the continuous variation in velocity, and the analysis of finite motion as the progressive summation of *conatus* gives warning that it is impossible *a priori* to find the dynamical principle of that finite motion in a simple proportionality.[1]

Then again, Leibniz knew, having learned it long since from Huygens, that the quantity of motion in the Cartesian sense (where velocity is considered only in respect of its magnitude, without taking into account either the sense or the direction) is not conserved in the phenomenon of impact, but only the quantity of algebraic motion. He learned also the part played by elasticity and so arrived at an explanation of the true laws of impact; finally, he understood the conservation of mv^2. Consequently, he had at hand the elements of a new system.

This system is characterized by the fact that the addition of the concept of mass to extension is not even sufficient to account for mechanical actions and motion. The concept of force is necessary in addition, and that concept is a distinct, dynamic reality. It does not depend on mathematical tricks; it is a primary concept. Its measure does not depend on mathematical analysis; it is to be discovered in reality through that which is provided by nature.[2]

In this perspective, the idea that Leibniz had already maintained of a universal elasticity acquires a new value and, furthermore, is recorded against the Cartesian thesis. By virtue of the principle of continuity, we must imagine in fact a gradual loss and an equally gradual restoration of motion, in both cases very rapid, during the impact of two bodies. That is easily explained by elasticity or intrinsic conserving force. The world of perfectly hard, homogeneous, geometric bodies of Cartesian mechanics is consequently replaced by a universe ruled by motion, the generator of matter itself, together with elasticity, the guarantor of conservation: in short, it is the (kinetic) energy which is conserved and not the quantity of motion.[3]

1. See M. Guéroult, *Dynamique et Métaphysique Leibnizienne*, Paris, Vrin, 1934, pp. 44–45.
2. See *v. infra*, pp. 52, 101-102.
3. See M. Guéroult, *Dynamique et Métaphysique Leibnizienne*, Paris, Vrin, 1934, p. 45; A. Hannequin, *La première philosophie de Leibniz*, pp. 123–124.

The principal work which opened the real attack of Leibniz on Cartesian mechanics is the *Brevis Demonstratio* ... published in the *Acta Eruditorum* (1686). The ensuing controversy, which we shall have occasion to consider in detail later, was carried on both as a refutation of the principle of the conservation of the quantity of motion by *reductio ad absurdum*, and as a criticism by default of Cartesian principles. In particular, if the Cartesian laws of impact are *irreparably* false, it is because they violate the principle of continuity;[1] that had not been understood by Huygens,[2] and Leibniz had tried to convince Malebranche of the fact. Those laws sometimes have the result of making an infinitesimal change in the data correspond to totally different aspects of the phenomenon; and that is inadmissible.

However, it is easier to follow a proof by *reductio ad absurdum* than to grasp what is at stake in the discussion it concerns. In order to understand the proof and the defeat of the adversary some patience and attention are needed. What is at stake in the discussion demands that vision of the whole which we have tried to make clear and it concerns something remarkable: a rationalism in quest of principles adequate for a world system.

Infinitesimal analysis and realism of force constitute its structure. It remains to be seen in the detail of the texts of limited objective, such as we are considering here, how the coherence of a thought is put solidly to the test in history.

1. *Nouvelles de la république des lettres*, February 1687, p. 139.
2. For Huygens, the Cartesian laws of impact are false, but from the point of view of a calculator who does not go to the root of things.—Letter from Huygens to Leibniz, 11 July 1692. Gerhardt: *Briefw. v. Leibniz mit Math.*, Vol. I, p. 699.
   See also Pierre Costabel, 'La septième règle du choc élastique de Christian Huygens', in *Revue d'Histoire des Sciences*, 1957, X. fasc. 2, p. 120.

I

From the history of a discovery to the discovery of history

Discovery of two Leibnizian manuscripts

Register No. 13 of the manuscript reports of the *Académie Royale des Sciences* at Paris mentions without any other particulars that on Wednesday, 26 March 1692, M. de La Hire put forward a paper by 'M. Leibniz' and 'some difficulties which he will explain at the first opportunity', and that on Saturday, 28 June of the same year, M. de La Hire read the '*Élémens dynamiques de M. Leibniz avec ses remarques*'.

Our attention had long been drawn to the possible interest of the Leibnizian paper read to the *Académie* by La Hire, but the registers of the manuscript reports and the publications of the *Histoire* and the *Mémoires de l'Académie* contained no reference to this matter, so we concluded that it was impossible to profit from the very brief indications which had been brought to light.

In connection with another investigation, we were turning over some bundles of anonymous and unclassified, miscellaneous papers in the archives of the *Académie*, and discovered, at the beginning of February 1956, a notebook of fourteen pages (19 cm × 28 cm) with a loose sheet of the same size inside. This discovery completely altered the situation.

The first twelve pages of the notebook are numbered and written on both sides. They give a copy of a memoir written by Leibniz, which was sent to Pelisson, 8 January 1692, and published in 1859[1] by A. Foucher de Careil in the appendix to the third edition of Letters of Leibniz under the title of *Essay de Dynamique*. The notebook has no title, nor any indication of author, nor is it signed. However, even before the identity of the text with that of the *Essay de Dynamique* had

---

1. We have to thank Professor J. E. Hoffmann for information which enabled this identification to be made quickly. See volume I of *Œuvres de Leibniz*, published by Foucher de Careil between 1859 and 1875. This volume was republished in 1867.

been established, we had had no trouble in discovering its author. The main topics in the controversy opened by Leibniz with the Cartesians in 1686,[1] and the typically Leibnizian expressions are found again in the above item with a clearness which, *a priori*, admits of no doubt. On the other hand, the loose leaf, mentioned above, which is inserted between pages 2 and 3 of the notebook, provides additional important arguments for the identification. Written on both sides in the same hand as the notebook, that page reproduces with variations a text published in the *Journal des Sçavans*, 7 September 1693, under the title *Règle générale de la Composition des mouvemens par M. d. L.* The same title is in fact placed at the head of the copy with which we are concerned. Examination by transmitted light proves up to the hilt that this copy was made under the same conditions as the notebook in which it is inserted, and that it has had the same fate. There is the same watermark, the same condition of the paper, and in particular the same traces of dampness. Finally, internal evidence makes it impossible to place the date of writing before December 1692–January 1693.

Furthermore, there was no mistaking the origin of the two manuscripts. The writing not being that of Leibniz, we were dealing with a copy of two texts by that author from the period 1692–1693.

Now, the text of the notebook provides a didactic restatement of matters. It opens with three definitions, accompanied by scholia, and followed by two axioms and two 'enquiries', and comprises subsequently nine propositions with their proofs. The very style of this writing well deserves the title of 'Elements'. Finally, the last four pages, where the Leibnizian identity is even more accentuated, if that be possible, are entitled '*Remarques*'. It would seem that there is no doubt but that the contents of this notebook comprise that which La Hire read to the *Académie* on 28 June 1692.

Identification of the copyist. Character of Des Billettes

Our investigation would have been incomplete if we had not had the good fortune to be able easily to identify the copyist of the two manuscripts. It happens that the previously mentioned register No. 13 of the reports of the *Académie* has at the end a manuscript notebook

---

1. *Acta Eruditorum*, March 1686. *Nouvelles de la République des Lettres*, 1686–1687.

written in the same hand as the two documents in question. This note-book reports the meetings held between 1692 and 1696 of the commission appointed by the *Académie* to prepare the publication of a universal description of Arts and Crafts, and the secretary of that commission, the writer of the reports, was M. des Billettes.[1]

'He was well versed in history, the genealogies of the great families of France, even in the knowledge of books which is a subject apart', said Fontenelle in his *éloge* of Des Billettes. He added: 'He was deeply versed in the Arts, that prodigious number of curious industries unknown to all those who are not engaged in them, ignored by those who are engaged in them, neglected by the most well-informed scholars. . . . As the *Académie* had formed the design of preparing a description of them, it considered Des Billettes necessary for that purpose and chose him as one of its *pensionnaires mécaniciens* on its re-organization in 1699.' It will be noted that Fontenelle's account is not quite correct, seeing that Des Billettes had received this commission of trust relative to Arts and Crafts in 1692. On the other hand, the character sketch given by the amazing perpetual secretary of the *Académie* is quite accurate. The library of the *Institut de France* has amongst its manuscripts (No. 1557) a register of the '*dépôt de l'Académie Royale des Sciences au Louvre*' entitled *Adversaria de rebus probatis, pour les Sciences et pour les Arts*, which is entirely written in the hand of Des Billettes and which is the most unusual collection imaginable of various recipes: pharmacy, medicine, cookery, metal working, etc. Des Billettes certainly seems to have been one of those individuals who are interested in everything, urged by the collector's passion, garnering everywhere a little of the most varied information; in short, a man whom Leibniz, the friend of antiquaries[2] and eager for all the curiosities of knowledge, must have met in his travels.

If we may believe Fontenelle, who records a curious anecdote, Des Billettes was also extremely modest and showed a concern for the common good of humanity in the most insignificant details of his behaviour. This feature of his character, which was greatly to the

1. Gilles Filleau des Billettes, born at Poitiers in 1634; *pensionnaire mécanicien de l'Académie* (first holder), 28 January 1699; *pensionnaire vétéran*, 21 August 1715; died at Paris, 15 August 1720. (See Fontenelle, *Éloges*, vol. II, pp. 60–62.)

2. We need quote only one piece of contemporary evidence of this well-known fact. It is provided by the postscript to the letter from Malebranche to Leibniz, dated 8 December 1692, and refers to M. Toinard, antiquary of Orleans with whom Leibniz was in sufficiently regular contact to be able to use him as a go-between in his correspondence with France (*Journal des Savants*, 1844, p. 545).

honour of its owner, was certainly one to attract the attention of Leibniz.

In fact, the correspondence of Malebranche with Leibniz in 1679 shows that the latter had made the acquaintance of Des Billettes at the time of his stay in Paris during 1672 to 1676, and that he had retained a real attraction towards him. The fact that Des Billettes was attached to the person of the Duke de Roannez[1] was not immaterial to that meeting between them, but would not in itself explain the feelings expressed by Leibniz himself.

Indeed, on 22 June 1679, he wrote to Malebranche: 'If M. des Billettes is at Paris, and if you see him, I beg you to have the kindness to assure him that his illness has grieved me. I hope that it will pass and will not return. For the public must be interested in keeping individuals who can be so useful to it as he.'[2]

This certificate of usefulness, delivered by Leibniz to him whom we have discovered was the copyist of the texts of 1692 on mechanics, is rather moving. Nothing that Des Billettes had amassed on the many papers covered with his clear, fine and distinguished writing has been published. If he had been useful in the public good, as Leibniz says, it was undoubtedly because he was one of those rather self-effacing and modest workers who employ their strength in the service of others.

One passage in the register *Adversaria de rebus probatis*, mentioned above, completes the portrait of this scrupulous copyist. On page 144 we read: 'I had got to this place when I was informed that there is a printed memoir similar to the one I had started to abstract', and there the copy stops. Des Billettes was careful not to clutter the archives of the *Académie* with copies of memoirs published elsewhere. Seeing that he took the trouble to copy the *Élémens dynamiques* by Leibniz, we may be sure, *a priori*, that he did so solely on account of the rarity of that text.

However, it happens that we are not reduced to pure conjecture as to the circumstances and motives for making that copy. In order that the facts relating to those circumstances and motives may have their full significance, it is necessary, before going any further, to widen our outlook and to try and restore the framework into which the texts we have discovered naturally fall; that is to say, we must con-

1. See *Journal des Savants*, 1844, p. 500. Note by V. Cousin. It was through the Duke de Roannez that Leibniz received the *Pensées* of Pascal.
2. See *Journal des Savants*, 1844, p. 507.

sider the history of the battle Leibniz waged against Cartesian mechanics.

Leibniz and dynamics

Leibniz had announced as early as 1690 that he would produce a work on dynamics,[1] and we know from his correspondence that, having started to put his ideas in order at Rome in 1689,[2] after discussions with Auzout, an influential member of the *Académie Royale des Sciences* at Paris, he left his manuscript with his compatriot von Bodenhausen on passing through Florence on his way back to Hanover. The reason for doing so was that von Bodenhausen, preceptor to the Duke of Tuscany and a mathematician, had undertaken to 'sort out the text', to 'put it in order and even publish it'. 'I have only to send the final part of it', stated Leibniz to Foucher (May 1692),[3] 'but every time I think about it, a host of new things comes to me, so that I have not yet had leisure to digest them.' The situation was still the same in 1696, according to a letter to Johann Bernoulli dated 8 March; Leibniz was still not ready to write the final portion of the treatise he had left with von Bodenhausen with a view to publication. He continued to be beset on the one hand by the difficulty of expressing everything he had in mind on the subject, and on the other hand by the desire to be as exhaustive as possible.

However, being pressed by his friends, he consented to 'put something' of his 'thoughts on dynamics' in the *Acta Eruditorum* for April 1695.[4] This was the first part, which ought to have been

1. *Acta Eruditorum*, May 1690.
2.(*a*) Letter from Leibniz to Huygens, 15/25 July 1690, *Der Briefwechsel von G. W. Leibniz mit Mathematikern*, ed. Gerhardt, vol. I, p. 594: '*Il n'y a que cinq ou six semaines que je suis de retour de l'Allemagne et de l'Italie pour chercher des monuments historiques par ordre de Mgr. le Duc. J'ai trouvé bien peu de personnes avec qui on puisse parler de ce qui passe l'ordinaire en physique et en mathématiques. M. Auzout que j'ai trouvé à Rome nous promet une nouvelle édition de Vitruve.*'
(*b*) Letter from Leibniz to Johann Bernoulli, 8/18 March 1696, *Die philosophischen Schriften von G. W. Leibniz*, ed. Gerhardt, vol. IV, p. 412 and note 2. According to this letter, it was as a result of discussions with Auzout that Leibniz started to write at Rome a pamphlet '*in ordinem*' in which were proved '*toutes ces choses, de la force tant absolue que directive et du progrès du centre de gravité*'. He left the unfinished manuscript with von Bodenhausen, '*in mathematicis egregio*', on passing through Florence.
3. *Journal des Sçavans*, 2 June 1692. Foucher had written to Leibniz on 31 December 1691: '*M. Thevenot est fâché de ce que vous ne nous avez pas fait part de votre mécanique que vous avez laissée à Florence.*' See *Lettres et opuscules inédits de Leibniz*, ed. Foucher de Careil, 1854, p. 87.
4. Letter from Leibniz to Burnet, 11–21 June 1695, Dutens, vol. VI, p. 224.

followed by the second in May of the same year. However, the second part was never published. It was discovered amongst the manuscripts at Hanover, and together with the first part appears in volume VI of *Leibnizens mathematische Schriften* edited by Gerhardt; the two together, under the title *Specimen Dynamicum*, constitute a small treatise which shows a rather unfinished state of the mind of Leibniz as regards dynamics. The text that was put in order by von Bodenhausen and returned to Leibniz after the death of its editor, has, on the other hand, remained unpublished. Leibniz never finished it.

'That which is necessary is known and that which is more profound can be of use only to chosen minds. *Margaritae non sunt objiciendae porcis.*' No doubt, Leibniz, by expressing himself in that manner in his letter to Burnet dated 8 March 1695, had in mind the diffusion of his ideas on theological questions, but it is quite clear that if it were necessary for him to be reserved on that subject, then all reserve was not excluded in respect of dynamics. There was formidable resistance to be attacked and, if possible, overcome, particularly in France.

Gerhardt has published in volume VI of the work mentioned above an *Essay de dynamique* which was written at a later date because it makes reference to the conversion of Malebranche in 1698; the following passage taken from it is characteristic.[1] Leibniz says: 'It is a long time since I corrected and rectified that doctrine of the conservation of the quantity of motion and put in its place the conservation of something else absolute, and precisely what was wanted, namely, the conservation of absolute energy. It is true that in general my arguments do not seem to have been adequately considered, nor has the beauty of what I have pointed out been understood, as I notice in everything that has been written in France or elsewhere on the laws of motion and mechanics, even after what I have written on dynamics. However, seeing that some of the most profound mathematicians, after many disputes, have come round to my opinion, I look forward in due course to general approval.' If the dawn of better days for Leibnizian thought did not break, according to his own opinion, until after the conversion of Malebranche, then we may conclude that at the beginning of the last decade in the century the enterprise of publishing against Cartesian mechanics must have appeared to him as being very risky. He could not afford to make a false step; it was

1. *Leibn. math. Schr.*, ed. Gerhardt, vol. VI, p. 215 *et seq.*

furthermore necessary, after the controversies with Malebranche and Catelan, to educate opinion.

The Leibniz–Pelisson correspondence

There will be no surprise at the thoughts that accompanied the *Essay de dynamique* which was sent to Pelisson on 8 January 1692.[1]

'... Seeing that you intend the matter shall be thoroughly studied and that the public shall be able to judge of it, I thought it would be opportune to put my thoughts on this subject in better order; that is what I have tried to do in the accompanying Essay on dynamics, in which I have taken the matter a trifle higher than I did in the papers that served in the dispute. I have done so all the more readily in that I have been able to have a better understanding through the dispute itself of the preconceptions that can be misused. *Many other things will appear in my work on dynamics*, as well to explain everything *a priori* as to show its use and application in the solution of particular cases, but I have taken from it only that which seems to me the easiest and most suitable for the purpose of explaining the general principle of the conservation of absolute energy. *I should like this essay to be examined by the Rev. Father Malebranche himself. For he would perhaps show it favour*, the more especially as he has already given signs of candour in this matter, or else he would indicate the points on which the cause could be better informed. *After that, we would let it be judged by competent persons and perhaps even by some of the Académie Royale des Sciences*. That famous Company has quite altered since I left France. . . . Consequently, I think I know only MM. Thevenot, Cassini, M. l'abbé Gallois and M. du Hamel there..., I know M. Dodart and M. de La Hire by repute.'

We could not wish for clearer and more explicit direct evidence. The *Essay* sent to Pelisson was only an attempt, reduced to bare essentials, to gain the favour of the French *savants*. The complete treatise on dynamics would contain many other things, but first of all it was necessary to break down the resistance that had been revealed. If Malebranche agreed to examine the memoir, 'perhaps' he would give in; and we are conscious of the extent to which this 'perhaps' is charged with hope. It was necessary also to reach some 'competent persons' and members of the *Académie*.

1. A. Foucher de Careil, *Œuvres de Leibniz* . . ., vol. I, 1859, p. 237 *et seq.*

In a reply dated 19 February 1692,[1] Pelisson stresses the fact that, being well satisfied with the *Essay de Dynamique,* he had a copy of it made, and that M. Dodart has taken the original of the *first fragment* to the *Académie.* He adds that M. Pirot also has a copy of it. Our discovery confirms these facts. The manuscript by Des Billettes belongs to a group of copies thanks to which it was possible to realize the attack on the Parisian resistance according to the wish of Leibniz.

Before pursuing the fate that awaited this attempt, some details need to be emphasized. Pelisson had not received the original, the true autograph manuscript, of the *Essay de Dynamique.* This manuscript is still at Hanover, and Leibniz had merely had a copy made for sending to France. That fact is known from an undated letter which must be ascribed to the end of January 1692.[2] 'Casting my eyes on my draft of the *Essay de Dynamique,* I noticed that I had not supplied the figure for which the copyist had left a blank space. That is why I now send it to you. Furthermore, it has made me remember the cause of that omission. The reason is that I had re-read the beginning of the copy and had forgotten to read the remainder of it, having had my attention diverted and the courier having arrived, I did not remember the omission. That is therefore the cause of a mistake which will be found in the copy and which I beg of you to have corrected. In the last Scholium, a proposition is inserted and is marked by a line having underneath the words commencing *"deux corps se choquant directement . . .,* etc.": I find that *"la raison des vitesses avant le choc est réciproque de la raison des vitesses après le choc"* must be replaced by *"la différence des vitesses avant le choc est égale à la differénce réciproque des vitesses après le choc".'*

The autograph manuscript at Hanover carries numerous corrections and erasures; the sentence to which the above remark refers forms part of a marginal addition; and, in spite of the difficulty in reading, it is easy to verify that Leibniz had indeed at first written '*la raison des vitesses,* etc.', which he subsequently corrected to read '*la différence des vitesses,* etc.', thereby eliminating a rather bad mistake.

On the back of Pelisson's reply to the communication from Leibniz of 18 January, which reply is dated 27 January, we find some manuscript notes by Leibniz which are the draft of a fresh letter. 'Here is

---

1. A. Foucher de Careil, *ibid.,* p. 244.
2. The documents used here form part of the unpublished manuscripts at Hanover (LBH-MS., I., 19).

the figure which I had forgotten to include in my previous letter with a correction to my *Essay de Dynamique*.'

Finally, in his letter of 19 February, quoted above, Pelisson, in acknowledging receipt of a letter dated 4 February from Leibniz, definitely states that he had noticed the absence of the figure from the *Essay de Dynamique*, but that he was surprised not to find either the figure, or the correction of which Leibniz spoke in his last letter: either Leibniz had forgotten to put it all in the envelope, or Pelisson himself had burnt them accidentally with the envelope. It will be seen that the documents which remain leave some doubt on the matter. What is definite, is that Leibniz had allowed a copy of the *Essay de Dynamique* to go to Pelisson without the figure and containing a mistake, bad not only in itself, but also with respect to a principle which he was to assert a little later as being fundamental: namely, the conservation of relative velocity in the problem of impact. He very quickly had the concern to make the necessary corrections. As a result of some vicissitudes of correspondence, the corrections were still not in Pelisson's possession on 19 February—that is something which it is difficult to ascertain. However, as the copy by Des Billettes contains both the figure of the autograph manuscript by Leibniz as well as the exact wording of the addition relating to the impact of two bodies, it is certain, on the one hand, that Pelisson finally received the corrections as Leibniz wished; and, on the other hand, that Des Billettes worked with documents that had been brought up to date for the end in view.

One further remark. This story provides evidence of the rapidity with which Leibniz was obliged to work. He did not have time to review carefully the copies that he had had made; he was in a hurry through the sending of couriers and was in the midst of a heap of correspondence; either he did not always send that which he had the intention of writing, or he was guilty of forgetfulness and mistakes.[1]

It is now time to take up the thread of our investigation again: how was the scheme put forward by Leibniz in his communication to Pelisson realized?

1. See the notes to the text printed in the appendix, pp. 115–123. We have already had occasion to point out a mistake made by Leibniz having the same origin, see Pierre Costabel, '*Deux inédits de la correspondance indirecte Leibniz-Reyneau*' in *Revue d'Histoire des Sciences*, 1949, **2**, 329–332.

Circumstances of and grounds for making the copies

On 1 February 1692 Pelisson wrote to Leibniz:[1] 'I am not a visitor at the house of Father Malebranche, but I will arrange for your communication to be brought to him.. for I agree with you that his assent would greatly enlighten the truth and that his opposition would even contribute to it.' On 10 April 1692[2] he announced to his illustrious correspondent that six copies of his book on religious tolerance[3] had been communicated to the members of the *Académie* through the good offices of M. Annisson, director of the *Imprimerie Royale*,' so that they may see in the additions what you have done me the honour to write to me on the doctrine of Descartes, and so that they may be thereby prepared to see your *Élémens de la Dynamique*, telling them that I had them put in a fine, large hand, because you had ordered me to make them available for their judgement'. He added: 'I shall have them communicated to Father Malebranche by one of my friends who is one of his, too. He is the one whom I have employed for the French translations which are in that volume and in the preceding ones; his name is M. des Billettes, who claims to have seen and known you at Paris and who speaks of you with all the regard that he ought. I do not know if you remember him.'

A marginal note by Leibniz on Pelisson's letter shows that he had at least the intention of immediately writing to Des Billettes. It is not possible to say if he did indeed do so. The earliest letter from Des Billettes to Leibniz that we have been able to find is dated 16 November 1692,[4] and is in reply to one from Leibniz sent at the beginning of, or during, the summer. It confirms a letter from Leibniz to Pelisson, dated 18–28 October: 'Allow me also to take the liberty of asking news of M. des Billettes of whom I had made some requests on matters of interest in science so as not to write him an empty letter.' Des Billettes apologized on 16 November for his tardiness in replying to those enquiries on matters of interest, but not a word is said about the *Essay de Dynamique*. Either the matter already belonged to the past, or else Leibniz was being extremely careful in his *captatio bene-*

---

1. A. Foucher de Careil, *Œuvres de Leibniz . . .*, vol. I, p. 245.
2. Manuscripts at Hanover, HS Theology I, XIX, 7, fol. 621.
3. *De la Tolérance des Religions, Lettres de M. Leibniz et réponses de M. Pelisson ou 4ᵉ partie des Réflexions sur les différents de la Religion*, Paris, Anisson, 1692.
4. This letter and the following one form part of the unpublished manuscripts at Hanover.

volentiae. The phrase '*pour ne lui pas écrire une lettre vide*' used in his letter certainly remains rather curious, and would call for further explanation, if it were possible.

In May 1692,[1] Pelisson revealed that he had not yet given the *Élémens de Dynamique* to the members of the *Académie*. He said, 'I want them to be a little more eager for them, as I think they will be after what has been published about them as well in our book as in the *Journal des Sçavans*.' It was only on 29 June 1692[2] that he positively stated: 'I have had your treatise on dynamics copied in a large, fine hand. I have put it into the hands of the Abbé Bignon who should present it to the *Académie des Sciences*. However, it was only some few days ago that I gave it to him and I have not yet had a reply . . . I have not thought it necessary yet to communicate that treatise to Father Malebranche. I shall do so now without fail and it will be by M. des Billettes himself who is one of his friends.' Finally, on 19 October 1692 Pelisson write:[3] '. . . I start with your Dynamics. I have had two fair copies of it made, one for the *Académie des Sciences*, where it will remain, the other for communication to Father Malebranche, but as he has been absent for several months, I have put this second copy into the hands of M. de La Loubère, your friend and mine, who has not yet returned it to me. . . .'

Let us bring together all the facts of this correspondence. Pelisson, who was not in direct contact with Malebranche, thought of Des Billettes for the purpose of making the link. Nevertheless, we do not believe that the copy which we have discovered can be identified with the copy '*en grand et beau caractère*' announced by Pelisson and put into the hands of the Abbé Bignon in June 1692. The handwriting of Des Billettes, as we have emphasized, is an elegant, neat hand, but it has nothing in common with that of the copies in '*grand et beau caractère*', that still exist in the archives of the *Académie*, of certain memoirs for the years with which we are concerned (1692–1693). However, it is of little moment. The positive fact, namely, that Des Billettes was charged by Pelisson (at least that was the intention) to act as intermediary with Malebranche, is sufficient to fix the origin of our manuscript. Pelisson must have spoken to Des Billettes of his project. Des Billettes entered into correspondence with Leibniz.

1. A. Foucher de Careil, *Œuvres de Leibniz* . . ., vol. I, p. 285.
2. A. Foucher de Careil, *ibid.*, vol. I, p. 287.
3. Manuscripts at Hanover HS Theology I, XIX, 7, fol. 574. A. Foucher de Careil, *Œuvres de Leibniz* . . ., vol. I, p. 322.

Everything conspired in favour of an integration of Des Billettes in realizing Leibniz's plan of attack.

We are prevented from being more categorical, because a rather curious attitude on the part of Pelisson emerges from the texts which we brought together. Generally speaking, he seems to have played the part of a temporizer in the business entrusted to him by Leibniz in that on several occasions he in fact postponed fulfilling the proposed transmittals either to the *Académie* or to Malebranche. According to the letter of 19 October, he waited until the latter was absent from Paris before seriously concerning himself with the copy that was intended for him. It seems difficult to accept that Des Billettes received a definite, clear commission from Pelisson on the subject of the *Essay de Dynamique* with respect to Father Malebranche.

Failure of the approach to Malebranche

Was contact definitely made with him? It would seem rather unlikely. The copy intended for him was in the hands of La Loubère at the end of October 1692, and Malebranche had not yet returned to Paris. His treatise on the *Lois de la communication des mouvements*, which had appeared several months before, had been sent to Leibniz by the intermediary of Toinard only on 6 October.[1] The manuscripts in the Leibniz archives at Hanover still contain the copy of this treatise with marginal notes by Leibniz.[2] It is clear that the latter was not satisfied, although Malebranche had acknowledged what he owed to him. Leibniz immediately drew up a series of '*Remarques*' which he passed to Malebranche through the intermediary of Toinard and of Larroque, certainly during the month of November,[3] and it was to these '*Remarques*' that Malebranche replied on 8 December,[4] thereby resuming a direct correspondence with Leibniz which had been interrupted for several years. This letter of Malebranche's makes no reference either to the existence, or to the substance of the *Essay de Dynamique*. Finally, in the reply to this same letter, a reply which can be assigned only to the beginning of 1693,[5]

1. Manuscripts at Hanover BR Toinard, fol.. 7.
2. Manuscripts at Hanover LBH MS IV 325.
3. Manuscripts at Hanover BR Larroque, fol. 17–18 (draft), fol. 15–16 (copy). Gerhardt, *Die phil. Schr. v. Leibniz*, vol. I, pp. 346–349.
4. Manuscripts at Hanover BR Malebranche, fol. 19–20. Gerhardt, *Die phil. Schr. v. Leibniz*, vol. I, p. 343–346.
5. Manuscripts at Hanover BR Malebranche, fol. 21–22 (draft), fol. 23 (copy revised and completed by Leibniz). Gerhardt, *Die philos. Schriften von Leibniz*, vol. I, pp. 349–352.

Leibniz resumes the main argument of the *Essay* in very much the same form: 'It will never happen that nature substitutes one state in place of another, if they are not of equal force. And if the state *L* may be substituted in place of the state *M*, then reciprocally the state *M* must be able to be substituted in place of the state *L* without fear of perpetual motion.' It is quite obvious that Leibniz was not sure of having achieved by means of his *Essay* that by which he set such great store, and being in doubt he preferred to take advantage of the opportunity to return to the charge and to emphasize the most fundamental point.

Failure at the Académie Royale des Sciences

Was Leibniz any more fortunate with the *Académie*? He certainly reached that assembly. Pelisson, though he temporized, did in fact transmit the communication Leibniz had sent him, and his evidence is confirmed by the minutes of the meetings of 26 March and 28 June 1692 which we mentioned at the beginning. The convergence of dates and expressions leaves no doubt as to the identification of the text of the *Essay* with that read by La Hire at the meeting on 28 June.

The minutes of the meetings after that date are completely silent on the text, and it was not until Saturday, 14 March 1693, that we find Pierre Varignon presenting to the *Académie* an 'investigation of the argument by which M. Leibniz claims to prove that God does not conserve the same quantity of motion in the universe'. Varignon's memoir, still unpublished, is in the archives of the *Académie*. It consists entirely of a criticism of the article by Leibniz in the *Acta Eruditorum* for March 1686, and makes no reference to the *Essay* of 1692.

The matter is partly explained because we know through Foucher that the 'dynamics' of Leibniz was *under seal*[1] amongst Thevenot's papers at the same date (March 1693). Thevenot had died on 29 October 1692,[2] and his successor was apparently decided upon in March 1693; so there is no need to assume ill-will in respect of

1. Letter from Foucher to Leibniz, 12 March 1693. A. Foucher de Careil, *Lettres et opuscules inédits de Leibniz*, 1854, p. 104.
2. Melchisedech Thevenot, who was concerned with the founding of the *Académie des Sciences*, having thereby ensured a continuation of the scholarly meetings at de Montmort's house, had been made keeper of the king's library in 1684. He took pleasure in collecting books on all kinds of subjects, principally philosophy, mathematics, etc. He had great influence within the *Académie* and was in direct relations with Leibniz.

Leibniz in order to account for the anomaly found in the form of Varignon's memoir.

Moreover, the latter had received official intimation from the *Académie* about the document of Leibniz at the beginning of July 1693, and was commissioned to refute it. We know this again through Foucher.[1] The refutation never saw the light of day. Varignon possibly considered it to be pointless after his intervention of the 14th March. It is indeed strange to find that fifteen years later, on 28 April 1708,[2] Varignon, when writing to Leibniz on the same subject and recalling the '*remarques*' that he had formerly written, again refers only to the article in the *Acta Eruditorum*. He had, therefore, completely forgotten the very existence of the *Élémens dynamiques*, the text, and perhaps the original copy, of which had nevertheless been entrusted to him in July 1693. It is true that in 1708 Varignon was recovering from a long illness caused by excessive intellectual work, and that his memory might have been enfeebled in consequence. It is also true that he was not particularly orderly in caring for his papers, seeing that he declared that he was unable to find his own manuscript. However, if it be reasonable to assume that all memory of the memoir from Leibniz in 1692 had disappeared, the fact remains that from the outset, and again after consideration in July 1693, Varignon judged it to be without major interest with respect to earlier printed texts on the same problem.

Whether it be for the same reason, or from deeper motives, the fact still remains that the purpose pursued by Leibniz was not realized, and that the attitude of the *Académie* remained very reserved.

As we have already said, Pelisson had written in May 1692: 'I have not yet given your *Élémens de la Dynamique* to the members of the *Académie des Sciences. I want them to be a little more eager for them*, as I think they will be after what has been published about them as well in our book as in the *Journal des Sçavans*.' On 30 March 1693,[3] Foucher explained: 'I do not have your dynamics. *It is better to send printed copies rather than manuscripts, for the printed copies can be communicated to several individuals and are preferred by some*.' It is possible that the method adopted by Leibniz, namely, the distribution of copies of the same manuscript, or of several closely related

1. Letter from Foucher to Leibniz, 15 July 1693 (A. Foucher de Careil, *Lettres et opuscules . . .*, 1854, p. 116).
2. Gerhardt, *Leibn. math. Schr.*, vol. IV, p. 166.
3. A. Foucher de Careil, *Lettres et opuscules . . .*, 1854, p. 111.

manuscripts, could indeed have been bad tactics, in that it did not allow his friends or sympathizers to express their approval publicly. However, it is definite that Leibniz was under no illusions in other respects regarding the grounds of the difficulty.

He wrote to Pelisson on 3 July 1692,[1] saying: 'Your opinion is right, that the members of the *Académie Royale des Sciences* must not be pestered *if it does not appear that they are pleased to see that which has been prepared for them.* Those are fruits that grow best on their own soil, which is so well cultivated under the protection of one of the greatest kings that has ever been.' In his reply of 3 August 1693,[2] Foucher said: 'We are very glad *that you give a rational interpretation[3] to the doubts of the Academicians.* It is the best apology you can make for them. I shall be enchanted to see their opinions directed and enlightened one day through your good offices; but you will be obliged from time to time to lend them some light of your understanding as you have made a beginning.' We may excuse him for expressing himself with such haughtiness and rather biting, chilly irony. To the extent that the *Académie* remained silent on the efforts of Leibniz to renew the discussion on mechanics, this masked resistance could only prove irritating for the great German philosopher.

It has been rightly said that he was always harping on the same subject. Certainly, when he had found the expressions that seemed to him the most striking for conveying his thought, he made very little change in them in the numerous texts, letters or memoirs in which he widely distributed it. For that reason it was rather easy to identify the manuscripts of which we have spoken. Consequently, must we consider that Varignon was not mistaken in having forgotten the memoir of 1692 and in having remembered only the contribution to the *Acta Eruditorum* of 1686, and that the two copies by Des Billettes can tell us nothing further? The answer can be supplied only by examining and commenting upon the texts.

1. Letter from Leibniz to Pelisson, 3–13 July 1692. A. Foucher de Careil, *Œuvres de Leibniz . . .*, vol. I, p. 297.
2. A. Foucher de Careil, *Lettres et opuscules . . .*, 1854, p. 119.
3. The rational interpretation mentioned here is that memoirs sent to the *Académie des Sciences* ought to be written in 'letter form', otherwise it is difficult to get them published! Letter from Foucher to Leibniz, 15 July 1693. A. Foucher de Careil, *Lettres et opuscules . . .*, 1854, p. 116.

II

The *Essay de Dynamique*

State of the text

Correction of Foucher de Careil's version

The notebook of Des Billettes is, as we have already mentioned, a faithful copy at second hand of the autograph manuscript of the *Essay de Dynamique*. It contains the figure which was forgotten by the first copyist at Hanover as well as the very important correction, relating to the conservation of relative velocity during the impact of two bodies, that Leibniz had made almost immediately after sending off the memoir.

The text published by Foucher de Careil in the appendix to his work of 1859 is not so accurate. In the scholium to axiom 1, we read: '*4 se peut substituer à la place de l'état M . . .*' whereas Des Billettes has, correctly: '*. . . si l'état* L *se peut substituer . . .*' In the *Remarques*, Foucher de Careil's text gives: '*. . . la force morte . . . a la même raison à l'égard de la force vive (qui est dans le mouvement même) que le point A à la ligne*'; and Des Billettes has: '*. . . que le point à la ligne*'.

The reader could easily correct the first slip himself and read the letter *L* instead of the figure *4*. When we examine the original manuscript, we find that the *L* written by Leibniz really has the shape of a *4*, so that it is quite easy to understand how a copyist who was not a specialist in mathematics could have made the mistake.

In the second case, the presence of the letter *A*, used elsewhere in the preceding part of the text to designate a material point, can only be explained by unseasonable initiative on the part of the copyist. It falsifies the true nature of the comparison desired by Leibniz, namely, in the same way that in geometry the line is a finite whole composed of infinitesimal elements called points, so motion in the finite state is produced by motions in the nascent or differential state.

Finally, Foucher de Careil's text has '*Remarque*' in the singular, where Leibniz wrote '*Remarques*' in the plural, as is correctly reproduced by Des Billettes in his transcript. A very minor point, no doubt, especially as it would be obvious on reading that there are in fact several remarks! But it is a detail that is not without importance for identification with the communication made by La Hire to the *Académie* on 28 June 1692.

In the second edition of volume I of the *Œuvres de Leibniz* in 1867, Foucher de Careil makes no correction to the various points that we have just stressed. He forgets the figure and the word '*réciproque*' which characterizes the equality of the differences between the velocities before and after impact. He has misread the original manuscript, if indeed he really used that document.

This statement about details which may seem to be of no great importance nevertheless draws attention to two particularly characteristic passages in the text which we reproduce fully, and the story of our investigation is consequently determined.

The controversy of 1687

We have mentioned above that the scholium to axiom 1 has exactly the substance as the reply given by Leibniz to Malebranche at the beginning of 1693.[1] Seeing that Malebranche had not replied in a general way to the arguments he had put forward regarding the laws of impact, Leibniz advanced the discussion.

He said: 'As for the laws of motion, we agree that the force is not lost, but that it is a question of knowing if that force which is conserved must be evaluated by the quantity of motion, as is commonly believed. The Abbé Catelan did not understand my view at all, and if he were my exponent to you, as it would seem, he could not have given you a good idea of it.'

The allusion is precise. Leibniz believed that Malebranche did nothing further with respect to his doctrine of mechanics after the stage to which it had been brought by the controversy with Catelan in the *Nouvelles de la République des Lettres* in 1686–1687.

Then, he has something better to say. He continued: 'In my view, if their force (from several equivalent bodies) were employed until exhausted in raising some heavy body, the effect would always be

---

1. Manuscripts at Hanover BR Malebranche, fol. 21–22 (draft), fol. 23 (copy revised and completed by Leibniz). Gerhardt, *Die phil. Schr. v. Leibniz* ... vol. I, pp. 349–352.

equivalent and would always reduce to raising the same weight to the same height, or to producing some other determined effect. However, I take weight as being the most convenient. That being accepted, I prove that the same quantity of motion is not conserved. I prove also that if two events, which according to the popular notion of force (= quantity of motion) are equivalent, follow one another, there would be perpetual mechanical motion. For example, if it should happen that all the force of a body *A*, of 4 pounds weight and 1 degree of velocity, were transferred to the body *B* of 1 pound weight, and that the body *B* should then receive, in the popular opinion, 4 degrees of velocity, I prove that we should unquestionably have perpetual motion.' Then follows the passage mentioned above which reproduces very closely the scholium of Axiom I of the *Essay*. However, what we have just transcribed has a very great importance here, for the plan of the *Essay* itself is given in several lines.

The plan is characterized by the part played by general ideas of equivalence and of the impossibility of perpetual motion for a definition of force, and in order to have a proper understanding of what is new in that outlook it is necessary to say something about the controversy of 1687.

In replying to the Abbé Catelan in the *Nouvelles de la République des Lettres* for February 1687, Leibniz had again used his familiar example and furthermore had stressed the absurd consequence of the Cartesian principle: If the force (regarded as a quantity of motion) of a body of 4 pounds and 1 degree of velocity be completely transferred to a body of 1 pound, the result must be 4 degrees of velocity, and as the weights are inversely proportional to the distances they can cover by virtue of their velocities, whereas the distances are as the squares of the velocities, it follows that the transfer of the force from a body of 4 pounds to a body of 1 pound allows the latter to go four times further than if the same force were applied to it (because, according to Descartes the same force is required to raise 4 pounds through one foot as 1 pound through 4 feet). '*Consequently, we should have drawn three times that force from nothing*' thanks to the transfer from one body to the other. 'That is why', said Leibniz, 'I believe that instead of the Cartesian principle, we ought to be able to establish another law of nature which I hold as being most universal and most inviolable, namely, that *there is always a perfect equivalence between the full cause and the whole effect*. It says not only that the effects are proportional to the causes, but in addition that each entire effect is equivalent to the cause. Although this axiom is quite metaphysical, it is nevertheless one of the most useful, and it provides the means of reducing the forces to geometrical calculation.'

Some would be able to find only the smallest difference between such a text and those of 1692. Leibniz expresses his thought there with a vigour that seems to leave no loophole for his opponent. The latter, however, did not consider himself beaten, seeing that in the *Nouvelles* for June 1687, Catelan commented: 'All the contradiction that M. Leibniz finds between Descartes and Galileo derives only from the fact that he is satisfied to judge forces by their effects, or motions by the distances covered, without having regard to their duration. . . . When there is equality between the quantities of motion or the forces and inequality between the bodies moved, the distances covered cannot be proportional to the reciprocal of the masses or proportional to the velocities except there be uniformity or equality of time.' The example given by Leibniz for the purpose of exposing the absurdity of the Cartesian principle is therefore unacceptable: 'M. Descartes speaks of mobile forces applied during equal times, Galileo compares forces applied, or motions acquired, during unequal times.'

As a matter of fact, the argument was not new, because it had already formed the substance of Catelan's first refutation (September 1686), to which Leibniz had replied in an article in February 1687 saying: 'M. l'abbé Catelan worries about the lengths of time during which the velocities are acquired' but 'the time is useless in that respect. . . . When there are two perfectly equal and similar bodies having the same velocity but acquired in one case as a result of impact, and in the other through a fall of considerable duration, shall we say that their forces are different?' That answer evidently did not satisfy the gainsayer; it set aside rather too rapidly an objection which was not lacking in pungency.

The proof of this pungency is that in the last article of the controversy (September 1687) Leibniz, after having taken up the previous arguments one by one in a clearer manner, and having asked for an answer to 'such precise and such easy things', contented himself with a contemptuous reference to the objection that had been made: 'I shall only add, as *hors-d'œuvre*, that I concede to M. l'Abbé that we can reckon force by time, though care is needed.'

Equality of the lengths of time being understood, Catelan saw no difference between the fact of transporting, with the same quantity of motion, a body 1 through a distance 4 and a body 4 through a distance 1, and of raising a weight 1 to a height 4 and a weight 4 to a height 1. That was a greater admission than Leibniz had hoped for in his interpretation of Cartesian thought, and confirmed him in the value of his argument. Indeed, after that, what is the use of being bothered by a consideration of the lengths of time! If the force of a body of 4 pounds and 1 degree of velocity be equal in value, according to Descartes, to the transport of a body of 1 pound through a distance of 4 in any direction, and if, on the other hand, we accept with Galileo that the distances covered by virtue of the velocities are proportional to the squares of the velocities, that is to say, 16 in the present instance for a body of 1 pound,[1] there is obviously a contradiction. It is a question of numbers, and we can understand the exclamation of Leibniz: 'If these misunderstandings happen in an argument that is almost entirely pure mathematics, what must we not expect in ethics and metaphysics.'[2]

We might wonder to hear Leibniz minimize a discussion in which he seems in fact to mis-handle the intention of Descartes on a point which is not pure mathematics. It has been pointed out several times, that in the line of argument adopted by Leibniz everything takes place as if he neglected the distinction carefully established (in Statics) by Descartes between force in 'two dimensions', composed of the weight and the distance covered, and force in one dimension. Descartes dispensed with a consideration of time and velocity when he stated the equivalence of 'force' 1 pound-4 feet and of 4 pounds-1 foot, and he was fully aware of the non-proportionality between velocities and distances covered by virtue of these velocities.[3] Catelan and those like him were therefore bad disciples and badly upheld the view of their master; they played him false by accepting that their opponent confuses the two Cartesian concepts of force as a quantity of motion and of force in two dimensions in order to elaborate the chief piece of a contradictory example.

Finally, what does the opponent himself think? Is he making game of disciples who are not fully conversant with their master's teaching and are unworthy of him? Otherwise, if he is not developing a clever argument *ad hominem*, is he deceived by the true Cartesian doctrine?

The numerical and contradictory example in question here had

1. Endowed with 4 degrees of velocity.
2. *Nouvelles de la République des Lettres*, September 1687, p. 952.
3. See P. Costabel, 'La démonstration cartésienne relative au centre d'équilibre de la balance,' in *Archives Internationales d'Histoire des Sciences*, April–June 1956.

already been the subject of the *Brevis Demonstratio*[1] with which Leibniz had opened the controversy in 1686. It was the great Arnauld who, in acknowledging the receipt of this '*petit imprimé*', had drawn the attention of Leibniz to the fact that the Cartesians had a very simple defence at their disposal.[2] 'I do not know if you have examined what M. Descartes says in his Letters about his general principle of mechanics. It seems to me that, in wanting to show why the same force can raise by means of a machine double or quadruple what it would raise without the machine, he states that he takes no account of the velocity' (28 September 1686).

Leibniz replied on 28 November 1686,[3] 'I have found in the Letters of Descartes what you have pointed out to me, namely, that he says there that he had purposely tried to cut out consideration of velocity in considering the grounds for common motive forces and to have taken only the height into consideration.'

This admission on the part of Leibniz was therefore after the event and thanks to Arnauld that he had direct knowledge of what Descartes meant. In the same letter, when pointing out that he had just examined Catelan's first refutation in the *Nouvelles de la République des Lettres*, he stresses that the latter is a wretched opponent and a bad Cartesian. 'He agrees with me more than I wish and he limits the Cartesian principle to the single case of isochronous forces ... which is quite contrary to the intention of M. Descartes.'

It seems, therefore, that Leibniz had been notified, right at the beginning of the controversy, of the essential point where his argument could be at fault; and yet, he modified nothing, substantially. The fact is he believed he could offer a technical objection to Arnauld's remark. 'It happens that he [Descartes] cut out a consideration of velocity where he could have kept it and he retained it in those cases where it gave rise to errors.' 'It happens' in fact in the case of statics and of 'potential force' that 'the velocities are proportional to the distances' because it is a question of considering 'the *first* efforts to fall without having acquired any impetuosity through continuation of the motion.' However, when we pass to dynamics and consider 'the absolute force of bodies which possess some impetuosity... the evaluation must be made through the cause or the effect, that is to

1. Gerhardt, *Leibn. math. Schr.*, VI, pp. 117–123.
2. Letter from Arnauld to Leibniz, 28 September 1686. Gerhardt, *Die phil. Schr. v. Leibniz*, I, pp. 67–68.
3. Letter from Leibniz to Arnauld, 28 November/8 December 1686. Gerhardt *Die phil. Schr. v. Leibniz*, I, p. 80.

say, by the distance that [the body] can travel in virtue of [its] velocity' and there is no longer proportionality between velocity and distance traversed.

Leibniz did not alter his opinion on the flaw in the Cartesian method seeing that in the manuscript of the *Essay de Dynamique* of 1698–1699,[1] he has noted in the margin: 'Consequently it is surprising that M. Descartes has so well avoided the danger of the rock by taking velocity for force in his small treatise on Statics or potential force, where there was no risk, having reduced everything to weights and distances when it was immaterial, and he abandoned distances for velocities in the case where the contrary must be done, that is to say, when it is a question of percussion or kinetic force which must be measured by weights and distances.' As a matter of fact, if Leibniz did not alter his opinion it was because he had already made up his mind even before having read Descartes. In fact, in the *Brevis Demonstratio* we read that the five simple machines confirm *a posteriori* the proposition that the same force is required to raise one pound by two feet as two pounds by one foot, and confirm it *tanquam hypothesis*. Another hypothesis would be that of evaluating the force by the mass and velocity, but if there be agreement between the two 'hypotheses' in statics, there is a 'divorce' in dynamics, *in potentiis vivis seu concepto impetu agentibus*.

Consequently, it is quite easy to note that Arnauld's remark merely reached a mind whose attitude had already been decided. No doubt, Leibniz had not read Descartes before writing the *Brevis Demonstratio*; he would not have failed to say so to Arnauld if he had done otherwise. The reading of Descartes, however, had left him well aware of his own position. It was quite obvious to him that he had replied in advance to the possible objection, and he satisfied himself by reviving a criticism which seemed to him to be faultless: the mistake of Descartes was his failure to state the impossibility of carrying over into dynamics the equivalence accepted in statics between distance and velocity for evaluating force.

The attitude of Leibniz had already been decided, as we said, at short notice. The fact is, from the time of the *Brevis Demonstratio*, he came to a conclusion which he was never to cease from asserting throughout the controversy. *In Universum potentia ab effectu aestimanda est.* The only sure principle that allows us to pass without error

---

1. Manuscripts at Hanover HS XXXV, IX, 3.

from the domain of statics to the domain of dynamics is the evaluation of the 'power' or of the 'force' by the effect. We shall revert shortly to this fundamental point. We merely note here that this fact explains, in our opinion, why Leibniz did not give close attention to the validity of the example which he gave in order to show the contradiction between Descartes and Galileo. To the two questions put forward above—was it a question of a clever argument *ad hominem* in which Leibniz makes game of his opponents, or whether he was deceived by the Cartesian doctrine?—there is no clear answer.

Leibniz by this time had a conception of force which appeared to him to be universal and valid in all cases. He took matters rather arrogantly without, moreover, being very aware of it, and he initiated a discussion on numbers in order to confuse those who remained prisoners to old formulas. It mattered little to him that the discussion was ambiguous. Nevertheless, it would be quite easy to raise the objection that when he states the equivalence of (4 pounds—1 degree of velocity—1 foot) and (1 pound—4 degrees of velocity—4 feet) as the expression of Cartesian meaning, so as to proclaim contradiction of Galileo (1 pound—4 degrees of velocity—16 feet), it is he who misrepresents the meaning in order better to combat the doctrine of his opponent. It would be quite easy to point out to him that the *two* Cartesian equivalences, namely: 4 pounds—1 degree of velocity, 1 pound—4 degrees of velocity on the one hand and 4 pounds—1 foot, 1 pound—4 feet on the other refer to *distinct* concepts of force, and consequently, in strict logic, his contradictory example proves nothing. However, Leibniz acted as if he were unaware of that *distinction*. The Cartesians enabled him to do so, and whilst he gave some credit to the meaning and intentions of Descartes, he ascribed to the latter a confusion of ideas; and for the reason that he himself could not conceive mechanics with a *plurality* of 'forces'.

Too much of a metaphysician to accept an elaboration based on such a plurality, he cast his own fundamental hypothetical statement of the problem on his opponent. Paradoxically, because of that, he was able finally to bring the discussion to a head on a point of 'pure mathematics', namely, on the opposition between 4 and 4^2. However, it is a sure proof, that when one engages in play with those worthy Cartesians, who concede more than is necessary to their opponents, then the play is subtle, and in the end one is taken in to some extent.

Consolidation of the Leibnizian concept

Such was the situation with respect to which the texts of 1692 show a very definite progress. We have already quoted the letter which accompanied the *Essay de Dynamique* sent to Pelisson. It is worthwhile repeating here one characteristic passage. Leibniz said to Pelisson: 'I thought it would be opportune to put my thoughts on this subject *in better order*; that is what I have tried to do in the accompanying Essay on dynamics, in which I have taken the matter *a trifle higher* than I did in the papers that served in the dispute. I have done so all the more readily in that I have been able to have a better understanding through the dispute itself of the preconceptions that can be misused.'

On the admission of Leibniz himself, the controversy with Catelan had proved useful. It had made him aware of the 'preconceptions' the sources of error. However, as he does not say what these preconceptions are, we must try and see if we cannot conjecture them from this fresh exposition in which matters are 'in better order' and taken 'a trifle higher'.

The *Essay de Dynamique* of 1692 opens with a *definition*, 'On equal, less, or greater force'. The commentaries to Axiom I and the letter to Malebranche quoted above explain it perfectly. To raise the same weight to the same height, or to produce *some other determined effect*, is the criterion for judging equal forces. To reject it opens the way to perpetual motion, which all responsible learned persons have long agreed to regard as an impossibility. For the substitution of one force by another equal force could then take place with a margin of possibility for ready action in creation *ex nihilo*. The metaphysics of motion, for it is definitely a question of metaphysics, submits two closely interdependent basic requirements: a principle of equivalence by substitution or by free transfer, and the impossibility of perpetual motion.

Leibniz was certainly right to say to Malebranche: 'As for the laws of motion, we agree that force is never lost, but it is a question of knowing if this force which is conserved ought to be evaluated by the quantity of motion.' In 1692 the view of Leibniz on dynamics had as the starting point for all elaboration something equivalent to what we should nowadays call an axiom-definition, namely, that in the domain of force there can be neither free gain nor loss.

Consequently, everything takes place as if Leibniz had understood the necessity for not leaving room for verbal confusions.

This state of awareness does not emerge clearly all at once with the texts of 1692. A trace of it is found in the small work *De causa gravitatis*[1] (*Acta Eruditorum*, May 1690): '*Sed ante omnia logomachiae excludenda occasio est . . ., et neque hanc ego libertatem cuiquam nego, quam mihimet concedi postulo . . . quod si ex surrogatione eorum tale absurdum, quale est motus perpetuus, oriri nequeat, vires ipsorum dicemus aequales. Hac definitione posita facile tanquam corollarium concedet Cl. objector eandem vim in corporibus conservari, seu eandem esse potentiam causae plenae et effectus integri. . . .*'

As early as 1690 Leibniz was therefore preoccupied with excluding 'before everything else' opportunities for play on words; and that is what he expresses in an almost identical manner in the Remarks to the *Essay*. 'If anyone wishes to give another meaning to force, as indeed some are rather accustomed to confuse it with quantity of motion, I do not want to argue about words and I leave to others the freedom that I take to *explain the terms*. It is enough if I am conceded that which is a fact in my opinion, namely, that what I call *force* is conserved and not that which others have called by that name. Because otherwise nature would not observe the law of equality between effect and cause, and would make an exchange between the states, one of which substituted for the other would be able to provide perpetual mechanical motion, that is to say, an effect greater than the cause.'

The text of 1692 confirms the awareness of Leibniz of the necessity he was under of building a logical structure in order to avoid a battle of words; nevertheless, it is easy to see some important difference with the *De causa gravitatis*, even though at first sight it may appear rather subtle.

When he undertook the writing of the *Essay* in 1692, Leibniz wanted to 'adapt himself more to the popular ideas' and to 'avoid metaphysical considerations of cause and effect, for in order to explain matters *a priori* it is necessary to evaluate the force by the quantity of the effect taken in a certain way which needs a little more attention in order to be understood' (Scholium to Definition 1). He clearly adopts, then, the point of view *a posteriori* which consists in defining the equivalence of force by the substitution of a state ($=$ one

---

1. Gerhardt, *Leibn. math. Schr.*, **VI**, p. 199.

or more bodies under certain conditions of position, motion, etc.) for another state without fear of perpetual motion, the concept of substitution being itself linked with 'the exclusion of all action from without', and then trying to find how the force may be expressed as a function of the effect or of the velocity. Only, he no longer says, as in 1690, that the conservation of force in every transformation exclusive of external action, or the exact equivalence between total cause and full effect, may be readily conceded, as a 'corollary', once the definition has been stated. It is an axiom there now, and we can quite see why. The definition allows the use of one word to characterize two 'states' which may be substituted one for the other under the conditions indicated: it says that these two equivalent states have the same 'force'; but it defines only ideas and words. It serves no purpose if it is not accompanied by a proposition of fact. In the *De causa gravitatis*, Leibniz had incorporated in the definition the impossibility of perpetual motion *tale absurdum*. . . . He effectively improved his account in 1692 and put things in better order by putting after the definition an axiom on the conservation of force, which is equivalent, as he himself points out, to a statement of the impossibility of perpetual motion. Thus, 'It will never happen', as he adds in the scholium to Axiom 1, 'that nature substitutes one state for another if they are not of equal force'. The definition, as stated, is therefore not void of sense, but the proposition that gives it factual existence is an axiom because there is no logical deduction which can lay down the true behaviour of 'nature', starting from prior hypothetical data.

This axiom, Leibniz repeats, is comparable with that which states that 'the whole is equal to all its parts taken together', the use of which is so widespread in 'geometry'. 'The one and the other provide the means of arriving at equations and an analytical method.' These are valuable announcements. They reveal that the penetrating mind of Leibniz had reflected a great deal on the metaphysics underlying the construction of a mathematical structure and on the value of the principles of conservation in order thereby to ensure foundations translatable into an adapted symbolism. How, then, do we arrive at 'equations and an analytical method'? Leibniz merely gives a suggestion in Axiom 2. He says, 'This axiom is conceded. Nevertheless, it could be proved by axiom 1 and otherwise. Without that it would be easy to obtain perpetual motion.'

Axiom 2 states that 'As much force is required to raise one pound to a height of four feet as to raise four pounds to a height of one foot', that is to say, we are definitely confronted by a mathematical equation. The axiom is conceded, because it is, formally, the principle universally accepted in all mechanics and especially in Cartesian statics. However, Leibniz is insistent in emphasizing that, from his point of view, it is not a question of a true axiom. To raise one pound through one foot is a *specific effect*; and its multiplication by four may be carried out in two ways, either by taking four pounds, or by raising through four feet. The whole, namely, four pounds through one foot, is equal in all its parts taken together, namely, to four times one pound through one foot or to one pound through four feet. Axiom 2 is a logical consequence of Axiom 1, and it is quite true that to reject it would inevitably lead to acceptance, with the difference between the multiplications by four of the same fixed effect, of the existence of mechanical actions 'derived from nothing', and consequently to acceptance of the possibility of perpetual motion.

It is essential to note that the first example of 'equation' obtained from Axiom 1 is not concerned with two 'states' (such as bodies, velocities), but with the effects produced by the 'expenditure' of the 'power to act' contained in those states. The fact is not stated there, but it is obviously understood. The letter to Malebranche mentioned above is definite proof of it.

In 1687, when replying to Catelan, Leibniz added 'a remark of consequence for metaphysics'.[1] 'I have proved that the force should not be evaluated by compounding the velocity and the bulk, but by the *future effect*', that is to say, not by the quantity of motion *mv*, but by the rising virtual power of the motion. Leibniz added further: 'However, it seems that the force or power is something real already and that the future effect is not. Whence it follows that it will be necessary to assume in bodies something different from bulk and velocity, unless we want to deny bodies all power of acting.' In order to emphasize the importance of that statement, Leibniz had had the last words printed in bold type. In the conclusion to the *Essay* of 1692, the reader will find the following explicit statements: 'That is all the more reasonable' (namely, the exclusion of a consideration of time and motion for evaluating the force) 'seeing that motion is a transitory thing which strictly does not exist because its parts are never all together. Thus, in addition to mass and change (which is motion), there is something else in corporeal nature: namely, *force*. We must

---

1. *Nouvelles de la République des Lettres*, February 1687, p. 141.

not be surprised if nature, that is to say, the supreme wisdom, establishes its laws on that which is the most real.'

All these quotations taken together enable us to form a more definite and exact idea of what Leibniz had in mind. From 1687 to 1692 his opinion had not changed as far as *realism* of the force was concerned. Matter is not extension, and there exists in bodies a dynamic reality, which is the power to act and 'already' exists, and which is even that which is 'most real' in corporeal nature. However, this superlative, which appears only in the text of 1692, marks the completion of an evolution.

We can certainly speak of the 'state' of a body or of a system of bodies having such and such adjuncts of relations with other bodies and of velocities, but 'motion is a transitory thing which strictly does not exist', it is an elusive reality and one that cannot, therefore, serve to evaluate the virtuality which is contained therein. Motion manifests the existence of that virtuality; but without it, there is nothing. 'It is force that is the cause of motion which really exists.' That is why force is 'more real' than anything else and also why it is the basis on which nature establishes her laws.

Consequently, we have a better understanding of the grounds of Axioms 1 and 2. The abstract condition for equality of force in two 'states' is the possibility of reciprocal substitution of one for the other without any external action, but as soon as it is a question of making manifest this equality it is necessary to pass from potentiality to act, that is to say, to consider the *whole effect* obtained as a result of the complete expenditure of the 'power to act'.

Such is 'the quantity of the effect' which 'needs a little more attention in order to be properly understood', and which could serve to evaluate the force *a priori* if there were not some difficulty in starting in that way a logical structure for the new dynamics. The method *a posteriori* has the advantage of allowing us to see things more clearly and not 'to be deluded' by fallacious simplifications. It is too easy to speak of the effect; what we have need of is inseparable from two qualifications: it is future, that is to say, it is registered in a duration which assigns a limit to it; and it is complete, that is to say, in that duration it is a totalization.

The Leibnizian categories

Having indicated the essential features that completion of the *Essay* of 1692 brought to bear on the Leibnizian account, we are now in a better position to continue our study of it.

Axiom 2 is followed by Postulate 1 relating to the possibility of transferring all the force from body *A* to body *B*. In the Scholium, this transfer is imagined to take place through a succession of impacts on intermediate bodies, or by other means of transmission. Leibniz said: 'It is of no consequence if that happens directly or indirectly, suddenly or successively, *provided that initially only the body* A *was in motion, and that finally only the body* B *is in motion.*' This extremely simple criterion is one consequence of Axiom 1, for if conditions are as stated '[body *B*] must of necessity have received all the force of body *A*', 'otherwise some part of it would be destroyed'.

The complete expenditure of the force of state '*A*', which finishes as it ought with this body in a state of rest, does not consist here in producing an 'effect', but in causing *B* to pass from an initial state of rest to a certain dynamic 'state' of the same virtuality.

Leibniz added: 'We can imagine a certain contrivance for carrying out these transfers of force, but even though we should not describe its construction, it is enough that there should be no impossibility in doing so, just as Archimedes took a right angle at the circumference of a circle without being able to construct it.'

The significance of this surprising analogy is provided by Postulate 2. 'Seeing that it is a question here of *the line of argument for the purpose of evaluating the grounds of the case and not the application*, we may regard the motion as taking place in a void, so that there would be no resistance from the medium', and we may imagine that 'the surfaces of the planes and spheres are perfectly smooth so that there would be no friction and so on.' It is accordingly postulated that 'external hindrances be excluded or neglected'. Thus, in the same way that the regular inscribed, or circumscribed, polygon gives an idea of the length of the circumference on going to the limit, so 'the contrivance' that we imagine for the transmission of motion gives an idea of perfect transmission in the limit with, we might say, a lubrification suppressing all resistance or intermediate causes of absorption of the effort.

It is convenient to recall here the distinctions between three methods recognized by Leibniz in 1671 in his *Theoria motus abstracti:* *'Geometrica, id est imaginaria sed exacta, mechanica, id est realis sed non exacta, et physica, id est realis et exacta'*. The geometrical method shows how bodies may be developed, it 'imagines' and subjects this imagination to the control of non-contradiction which then renders it exact. The mechanical method is real because it includes the technical processes of production, and is inexact because, according to the contriver, these processes are more or less perfect. The physical method is that of nature. It is rigorous and results in realities as does the technical method.

Leibniz had not entirely abandoned these categories in 1692. Nevertheless, it is noteworthy that the example given even in 1671 for the purpose of illustrating the geometrical method, namely, the segment of a straight line which produces a circle in bending itself 'through minima', *i.e.*, into infinitely small sides, serves here as it happens as an analogy for the purpose of justifying a principle in the new mechanics. This new mechanics does not cease to be *realistic*, but in the limit it can join with the geometrical method if we agree to line up the difficulties and concern ourselves in the first place with 'the line of argument for the purpose of evaluating the grounds of the case'. 'The application' will then come into its own. The advance which is thus shown in the conception is registered in the course of logical effort of which the *Essay* of 1692 is evidence.

We repeat, Leibniz had not entirely abandoned these categories. In this connection, another example is provided by the Scholium to Definition 3, which we have so far neglected and which relates to 'mechanical' perpetual motion. The text of this definition is not a model of clarity, but we can all the same explain it without falsifying it by saying mechanical perpetual motion is that of a 'machine' in which bodies finding themselves initially 'in a violent state and straining to get out of it' cause displacements such that 'after some time, everything is once more in a state not only just as violent as it was at the beginning, but even more so'. The mechanical perpetual motion 'that we postulate in vain', as Leibniz remarks, is therefore that of a machine which has within itself the cause and support of its motion and which not only consumes nothing, has no contribution from outside, but is yet able to perform some service or some extra work. The Scholium to the definition indicates how such a machine may be devised: weights being raised to a certain height endeavour to fall and force other weights to rise. 'However, *nature* is deceived in believing it possible to achieve its purpose, and *art arranges matters so well that at the end of a certain time it happens that there are just as many weights in the raised position as at the beginning and even more.*' Art, that is to say, the technique of building the machine, believes therefore that it is able to restore the machine to conditions of disequilibrium at least equal to those prevailing at the beginning. 'It is true', continues Leibniz, 'that *if we remove the accidental hindrances the falling bodies can ascend by themselves to the original height. And that is necessary*; otherwise the same force would not be conserved, and if the force decreases, the whole effect is not equivalent to the cause, but is less. We can say therefore that there is a *physical perpetual motion*, as would be the case of a perfectly free pendulum, but that pendulum will never go beyond the original height, and it will not even reach that height if it brings about or produces the least effect in its path, or if it overcomes the least obstacle. Otherwise that would be *mechanical perpetual motion*.'

The contrivance which consists in attaching a weight to the end of a string can certainly 'deceive' nature by obliging the falling weight to rise again, but the best that can happen is for the height of regain to be the same as the distance of fall; and for that to be possible, all resistance from the air or any friction throughout the path must be eliminated. The perfectly free pendulum is definitely endowed with perpetual motion, but that perpetual motion is ideal and in proportion to the perfect freedom of the pendulum, which itself is only an ideal, limiting and abstract state.

Leibniz applies the term physical to this motion, because, conformably to the categories given in the *Theoria motus abstracti*, he reserves this qualification for that which is both real and conformable to the exact grounds of the case. In that sense, the new dynamics, which he puts forward in the *Essay*, and the development of the principles of which we have followed in some detail, is a physical dynamics. There is no doubt but that Leibniz would not accept the term of mechanics, even rational.

All this now enables us to understand a detail to which we have not drawn attention in order not to complicate things.

Everywhere in the foregoing, when speaking of the impossibility of perpetual motion in our commentary on Leibniz, we ought to have said *mechanical perpetual motion*, as he himself does. Only that kind of perpetual motion is impossible, because any real machine, whatever it may be, if it is going to function endlessly throughout an uninterrupted succession of cycles, contains internal resistances which can never be overcome gratuitously, however little they be. It is only that kind of perpetual motion, if it existed, that would exhibit creation *ex nihilo* in each of the cycles of which it would be composed.

The logical coherence of the Essay *of 1692*

With the benefit from the above preliminary study we are at last ready to proceed to an examination of what fresh matters the *Essay de Dynamique* of 1692 introduces as compared with the *De causa gravitatis* of 1690, that is to say, the new proof of the contradictory example which had formed the substance of the controversy of 1687 and which aims at justifying the evaluation of force not by mv, but by mv^2.

By means of a balance with unequal arms, one of which is a little more than four times the other, Leibniz proves 'easily by ordinary statics' that body A of 1 pound falling from a height of 16 feet can raise a body of 4 pounds to a height of 'somewhat less than 4 feet' (Proposition 2). He then proves (Proposition 3) by assuming the quantity of motion to be always conserved, that we can arrange matters so that instead of a body of 4 pounds with one degree of velocity we have a body of one pound with 4 degrees of velocity. Finally (Proposition 4), a device which resumes the balance of Proposition 2 enables him to establish that we can obtain *mechanical* perpetual motion if we 'assume that instead of 4 pounds with one degree of velocity, we can acquire one pound with 4 degrees of velocity'. Thus, the hypothesis of the conservation of the quantity of motion is reduced to absurdity.

What a contrast between this profusion of precautions and the texts of 1687!

Leibniz does not talk of the equality of 'force' (1 pound—$4n$ feet) and (4 pounds—n feet). He considers a balance slightly out of equilibrium according to the principle of 'ordinary statics'. No one can deny that the movement of this balance leads to the result of Proposition 2, namely, that a body of 1 pound falling from a height of 16 feet can raise a body of 4 pounds to a height of almost 4 feet without any intervention except its own action.

That being so, if we assume on the one hand the possibility of ' *transferring all the force*' from a body A of 4 pounds to a body B of 1 pound (Postulate 1) in such a way that 'nothing accidental or external absorbs something of the force' (Postulate 2), body A being initially *alone* in motion, then body B, and if *on the other hand* we adopt the *hypothesis* that the quantity of motion 'if always conserved', the consequence of the transfer of force *and* of the hypothesis is that, if A had initially 1 degree of velocity, B receives 4 degrees of velocity.

Leibniz carefully avoids confusing the concepts of force and quantity of motion and applies himself to an exposition of his new line of argument with faultless logic. He asked (Postulates 1 and 2) that we accept the possibility of 'transferring' all the force of one body to another without loss or gain, and it is definitely a question of force in the sense as defined by him. The body A being initially alone in motion, then body B being alone in motion, it becomes obvious that if we adopt moreover the hypothesis that the quantity of motion, the product of mass and velocity, is 'always' conserved, then the stated conclusion must be recognized. Under the conditions laid down for the transfer of force, nothing 'accidental or external' intervenes, that is to say, the abstract system formed by the two bodies A and B being the only thing to consider, the conservation of the quantity of motion transferred from body A to body B results in effect from the general 'hypothesis' of the conservation of the quantity of motion.

As for the device indicated in Proposition 4, it does not call for any particular comment. Its internal logical coherence is complete when we accept the preceding two propositions. It suffices to give here a transcription of the proof.

'Let a sphere A of [4] pounds[1] weight fall from a height of one foot and acquire one degree of velocity. Now, instead, let a sphere B of one pound have 4 degrees of velocity by hypothesis. This sphere B will be able to rise to a height of 16 feet (Proposition 1) and then coming to an equilibrium which would be reached at the end of its rise and falling anew from that height to the ground, it will be able to raise A to a height somewhat less than 4 feet (Proposition 2). Now, at the start, the weight A was raised one foot above the ground and B was at rest on the ground. Now, it happens that B having fallen again is still at rest on the ground, but that A is raised nearly 4 feet above the ground . . ., thus A, before returning from the height of 4 feet to its original height of 1 foot, will be able to perform some mechanical effect as it goes along . . . and that activity will be able to continue for ever and that means achieving mechanical perpetual motion.'

---

1. See note to the text in the appendix, p. 120.

There is nothing to be retracted in that argument. To assume that the quantity of motion is 'always' conserved effectively results in the elaboration of a mechanism endowed with *mechanical* perpetual motion, that is to say, it leads to an absurd conclusion. Furthermore, as the hypothesis of the conservation of the quantity of motion has been applied to the case of a simple transfer of force, we must consequently definitely abandon the idea of evaluating force by the quantity of motion.

Propositions 6 and 7 that follow the argument which we have just been considering in detail are intended to show that the role which we sought in vain to be taken by the quantity of motion mv is certainly filled by the quantity mv^2.

These propositions do not call for any particular comments; they are logically linked with what precedes. Proposition 6 shows that the state '4 pounds, 1 degree of velocity' has only one quarter the force of the state '1 pound, 4 degrees of velocity', and that results from comparing the heights of rise which characterize the exhaustion or expenditure of the forces. Proposition 7 shows that the state '4 pounds, 1 degree of velocity' has the same force as the state '1 pound, 2 degrees of velocity', and the conclusion follows. 'It is worth noting', says Leibniz, 'that all these propositions and many things that are said here could be made known and stated more generally according to the manner of geometers. For example, we could say in general that *the forces of bodies are in a proportion compounded of the simple proportion of their mass and of the duplicate proportion of their velocity.*'

Thus, the method *a posteriori*, leaning on the definitions and axioms that we have already encountered, and based on the impossibility of mechanical perpetual motion, provides not only 'a short way for evaluating the effects by the forces, or the forces by the effects' (see *Remarques*), but also results in an expression for the force of a state as a function of its *actual* elements: mass and velocity.

In order to terminate the discussion, Leibniz draws attention to two points on which an opponent could still be capable of being mistaken.

In the first place, it is quite true that in the case of the impact of two bodies, there can be conservation of both the total quantity of motion and of the total force. However, that assumes that the difference between the velocities of the two bodies before impact is equal to the 'reciprocal difference' of the velocities after impact, that is to say, it assumes a perfectly elastic impact. Furthermore, the bodies must be moving in the same direction, both before and after impact. This precise detail on the part of Leibniz is perfectly true if we understand the quantity of motion in the Cartesian sense, that is to say, as an absolute value, and not in the modern sense of an algebraic quantity. $mV_0^2 + m'V_0'^2 = mV_1^2 + m'V_1'^2$ and $m.|V_0| + m'.|V_0'| = m.|V_1| + m'.|V_1'|$ are indeed simultaneously true if $(V_0 - V_0') = -(V_1 - V_1')$ and if V_0, V_0', V_1, V_1' are of the same sign. If this last condition is not satisfied, we have $mV_0 + m'V_0 = mV_1 + m'V_1'$, that is to say, that the quantity of motion is conserved only in the algebraic sense. That was what troubled Huygens.[1] On a somewhat lower plane, it was also what Malebranche had difficulty in understanding,[2] and the annotations by Leibniz in the treatise on *Lois de la Communication des Mouvements* emphasize the fact. The comment has some importance. In endeavouring to anticipate a possible objection, namely, 'what do you make of those cases in which, as in the theory of impacts, there is conservation of the quantity of motion', Leibniz gives proof of the extent he is master of the subject. He was capable, even if he had some doubts, as is shown by the first version he sent to Pelisson, of taking account of the very special conditions where there is conservation in the Cartesian sense. He definitely regarded them as very special conditions.

The second point to which we refer concerns the simile frequently used in mechanics towards the end of the seventeenth century in order to justify the conservation of mv. 'Mass [was likened] to water and velocity to salt that was made to dissolve in that water, and we may certainly imagine the salt being more extended in more water, or more confined in less water, and even withdrawn from one water and

---

1. As early as 1652, Huygens had proved that, in the impact of two bodies, the total quantity of motion is not conserved in the Cartesian sense. See Huygens, *Œuvres complètes*, XVI, p. 95, p. 167.

See also Proposition VI, ' *De motu corporum ex percussione* ', *ibid.*, XVI, pp. 49–50.

In 1654 he pointed out that the Cartesian formula could be true in the case where the velocities are in the same direction before and after impact, and he realized the possibility of conservation in the algebraic sense (see *ibid.*, XVI, p. 102), but he did not manage to reverse the roles and to consider the case in which the Cartesian principle succeeds, as a particular case.

2. Manuscript notes on the copy of *Lois de la communication des mouvements*, by the author of *Recherche de la Vérité*, Paris chez A. Pralard, November 1692. Manuscripts at Hanover LBH. MS. IV, 325. Malebranche makes a special case of 'contrary velocities', p. 27. 'That saves him', wrote Leibniz, 'from the objection I wanted to make, but the rule still holds even when the velocities are not contrary.'

transferred to another.[1] The intentions were good seeing that this flight of imagination had its source in the fact that 'it was not possible to understand how one part of the force could be lost without being utilized somehow, or gained unless coming from nowhere' and that was certainly what was underlying the thought of Leibniz, too. Yet, he was right in saying that by the simile in question 'there has been a transgression against true metaphysics and against the skill of evaluating things in general', for velocity and motion have been made identical, the first has been treated like matter, and the primary consideration has not been instituted on the basic principles and concepts as done by Leibniz and which is the only sure way of making progress. No doubt, Leibniz had pointed that out, but it is interesting that he should have reverted to the matter in the *Essay* of 1692, and that he should have left evidence of the clarity he had acquired on one of those more fallacious modes of thought which he seems to relate more closely to ordinary, natural concepts: dilution and homogeneous distribution.

It now remains to consider more closely this force which is evaluated by mv^2 and which Leibniz calls 'living'.

Kinetic force and potential force distinguished

Tradition ascribes the expression kinetic force to Leibniz. M. Guéroult has pointed out that the term '*vis viva*' makes its first appearance in the *Specimen dynamicum*[2] and notes that Leibniz more often than not used the expression '*potentia*' to designate mv^2. Nevertheless, it can be easily shown that the expression '*potentia viva*' in contradistinction to '*potentia mortua*' was already part of Leibnizian language in 1686.

In fact, apart from the lines that we have already quoted from the *Brevis demonstratio* and which are relevant here, there is the following most important statement: '*Est autem potentia viva ad mortuam vel impetus ad conatum ut linea ad punctum vel ut planum ad lineam.*'[3]

It is substantially the same as the passage in the '*Remarques*' at the

1. The same is formulated in the treatise of I. G. Pardies, S.J.: *La Statique ou la Science des forces mouvantes*, Paris, 1st edition, 1673. Leibniz had read Pardies; he had met him at Paris, and said: '*Fuit mihi cum Pardesio . . . consuetudo non vulgaris . . .*', Gerhardt, *Leibn. math. Schr.*, VI, p. 81.
2. M. Guéroult, *Dynamique et Métaphysique Leibniziennes*, Paris, 1934, p. 33, note 4.
3. Gerhardt, *Leibn. math. Schr.*, VI, p. 121.

end of the *Essay*: we drew attention at the beginning of this chapter to the mistake in the text as published by Foucher de Careil. 'Equilibrium must be considered as consisting in a simple effort (*conatus*) before motion, and that is what I call *the potential force* which has the same relationship with respect to *the kinetic force* (which is the actual motion) as a point to a line.'

We are not responsible for underlining the terms potential force and kinetic force. It is to be expected that Des Billettes, the careful copyist that he was, has merely transcribed marks of emphasis originating with Leibniz himself.

The identity between the stated relationship and the geometrical representation when we go from the text of the *Brevis demonstratio* to that of the *Essay* of 1692 shows that Leibniz had not changed his views on this fundamental point during the period under consideration. Undoubtedly the term '*potentia*' is better suited than that of '*vis*' to his concept of 'force', seeing that we know that the realism of the latter is attached to the virtuality contained in the 'state' of the body and that the manifestation or whole effect, which allows of an evaluation, assumes expenditure of the 'power to act'. However, it would be obviously incorrect to reject in 1695, with the *Specimen Dynamicum*, the appearance of the concept of kinetic force under the pretext that the expression '*vis viva*' was used there for the first time. The texts to which we have just referred prove once again that the concept of kinetic force, clearly conceived in contradistinction to potential force, was used as early as 1687 by Leibniz, and the *Essay* of 1692 provides evidence that as soon as Leibniz wanted to translate into French he used the expression '*force vive*' [kinetic force].

Though it does not appear in the controversy with Catelan, it seems that it did not appear before 1692 only because Leibniz had gone more thoroughly into the matter and had consolidated his thought in the meantime.

The commentary that follows the stated relationship referred to above must claim our attention. 'At the beginning of the fall', said Leibniz, 'when the motion is infinitesimal, the velocities or rather the *elements of velocity* are proportional to the downward fall, whereas after the *elevation*, when the force has become kinetic, the distances traversed in falling are proportional to the squares of the velocities.'

The term 'elevation', which evokes a picture of upward motion, is ambiguous here and could be misleading. It is clear that it must be understood here as the equivalent of our expression integration. The

61

whole structure of the thought makes a distinction between the nascent or differential state of motion and the completed or integral state of the same motion. It is that completion which causes the force to pass from the passive to the active state. It now remains to be seen if, in his desire to characterize the difference between the two states, Leibniz has not made a mistake. Is it true that the change of state corresponds to such a radically different law in the relationship between the 'distances traversed in falling' and the velocities? At the present time we should be tempted to say that the infinitesimal 'distances traversed in falling' are of the second order whereas the velocities are of the first order and that Leibniz was therefore quite wrong.

Now, Leibniz does not say that the 'distances' are proportional to the velocities but that the 'elements of the velocity' are proportional to the 'distances', that is to say, the very definition of velocity can only be applied to the beginning of motion when this is infinitesimal. It is true that the expression assumes a fixed reference for an interval of time, but in the realm of infinitesimals that condition can easily pass unnoticed or appear fulfilled. Clearly, it is very difficult to decide how Leibniz effectively regarded matters in that respect.

What is certain, when we compare his reply to Arnauld, quoted above,[1] with the present version, is that Leibniz regarded infinitesimal motion as uniform motion. In 1686 he said that we must pay regard 'to the initial efforts to fall *without having acquired any impetuosity for the purpose of continuing the motion*'. He says nothing further here except the brief statement: 'when the motion is infinitesimal'. The first text explains the second in the sense that we have just indicated, for not to have acquired impetuosity for the purpose of continuing the motion is the equivalent of uniformity and constancy of velocity. The second text shows that in 1692 Leibniz did not feel obliged to justify that uniformity other than by an appeal to the nature of the domain in which it occurs: in the nascent state, motion is uniform because it is infinitesimal.

In a similar manner, on the occasion in 1686 when Leibniz replied that when we consider 'the absolute force of bodies which possess *some impetuosity*, the evaluation (of the force) must be made through the cause or the effect, that is to say, by the distance the body can travel in virtue of its velocity', which distance is proportional to the

1. See p. 45–46.

square of the velocity, he says here: 'whereas *after the elevation, when the force has become kinetic*, the distances traversed in falling are proportional to the squares of the velocities'. The closeness of the texts is illuminating.

The 'impetuosity', namely the energy and its 'power to act', manifests itself on a finite thing. When bodies pass from rest to stable motion, when the velocities are no longer infinitesimal but finite, '*after elevation*', there is a *kinetic force* and the distances traversed are proportional to the squares of the velocities. Leibniz clearly does not concern himself directly with justifying that; he continues to refer to Galileo and Huygens for that result, but he has divested his conception of things lacking in foundation. 'Like a point to a line', everything is explained by the distinction between the infinitesimal and the finite. Leibniz certainly suspected it as early as 1686 in the *Brevis Demonstratio*, but he knows now that there is nothing else to say and that once more geometry traces the path of truth.

Defect of the Leibnizian concept

It is certainly possible to divest the mind of imaginings and confused ideas which have helped to give form and solidity to one's concepts, but there is some danger, because analysis of what is discarded in that way runs the risk of causing some shortcoming. This 'impetuosity' of the finite state of motion, is it its stable actuality or its tendency, its thrust towards a fresh state? Is it its actual velocity or its acceleration? It is obvious that Leibniz never asked himself the question, being wrapped in the seductive clarity of the distinction between the finite and the infinitesimal. Consequently, he was unable to complete the design that he contemplated.

Furthermore, his wish to put aside a consideration of time, in which he linked up again with Descartes, did not allow him to bring to the analysis of the concept of motion and of velocity the careful rigour which he had applied to the concept of force.

In the *Remarques* he says: 'It is yet appropriate to point out that force can be evaluated without taking time into consideration. For a given force can produce a certain limited effect which it will never exceed however much time be allowed. Whether a spring is let down suddenly or gradually, it will not raise a greater weight to the same height, nor the same weight to a greater height.'

The essential difficulty which he stressed in 1687 in a passage already quoted, and which was expressed again in a letter to Pelisson at the end of 1691, namely, that 'motion is something successive, which consequently never exists, any more than time, because all its parts never exist together'.[1] Leibniz thought that he had got round the matter, thanks to the realism of force 'which exists intact at every moment' and whose quantity is independent of the time for its expenditure. It was not by chance that he emphasized in the *Remarques* that his method is a '*voie abrégée*'. Unfortunately, through this very abridgment, it is full of danger.

Conclusion

In conclusion, we must emphasize what this method owes to Huygens. Leibniz himself, in the *Brevis Demonstratio*, has acknowledged the inspiration that he derived from reading the controversy between Huygens and Catelan between 1681 and 1684.[2] The close connection, thrown into relief during that controversy, between the principle that the centre of gravity of a body or of a system of heavy bodies cannot rise of its own accord and the impossibility of perpetual motion evidently provided him with a clue.[3] Moreover, he made no secret of the fact that Huygens had been his master in mechanics.

We shall then not be surprised to find him dealing tactfully with this master at the time when he was trying to prevail against the opinion of the French *savants*. On 1 April 1692,[4] he wrote to him as follows: 'Re-reading lately your explanation of gravity, I have noticed that you are in favour of the void and of atoms. I admit that I have some difficulty in understanding the reason for such infrangibility and I believe we must have recourse to a kind of perpetual miracle for that purpose. However, seeing that you have a fondness for approving such extraordinary things, it must be that you see some considerable

1. *De la Tolérance des Religions*, Paris, Annisson, 1692, pp. 9 to 14. Letter of Leibniz to Pelisson.
2. *Journal des Sçavans*, 1682, pp. 122, 200, 224; 1684, pp. 142, 225, 312, Huygens, *Œuvres complètes*, VIII, Nos. 2260, 2265, 2267; IX, pp. 80–81, 463. See *Brevis Demonstratio*. Gerhardt, *Leibn. math. Schr.*, VI, p. 120.
3. It is convenient to note here, that the impossibility of perpetual motion was also one of the main arguments of Pardies, whose works Leibniz had read and which certainly had an important influence on his thought. See I. G. Pardies, *La Statique . . .*, Paris, 1673.
4. Gerhardt, *Der Briefwechsel von G. W. Leibniz . . .*, Berlin, 1899, p. 694. Huygens, *Œuvres complètes*, X, p. 286.

justification in it.' Leibniz, who was then quite aware that 'nature does not operate by leaps', and that this 'axiom' 'destroys *atomos, quietulas, globulas secundi elementi,* and other similar impossible notions' and 'rectifies the laws of motion' (Leibniz to Foucher, January 1692),[1] Leibniz, who was aware of the gap separating him from his master, addressed him in a quite different tone from that which he used towards his Cartesian opponents.

Guéroult was quite right in saying that in making use of data supplied by Huygens, the philosophic spirit of Leibniz becomes apparent through the methodical synthesis of the great principles of conservation and through a systematic co-ordination of the various propositions, in which appear arguments with metaphysical repercussions that were neglected or not accepted by Huygens.[2] Furthermore, the characteristic of Leibnizian elaboration is the introduction of the vistas of the infinitesimal calculus and the clarity of the concept of absolute reality of force. Through attention to these two elements, considered philosophically, Leibniz definitely excels Huygens.

As we have already seen, his preference is given more to the realism of force. He discovered there a light which enchanted him and which extended far beyond the knowledge of motion. '*For I find nothing so capable of being apprehended as force*', he wrote to Bossuet in May 1694,[3] 'I believe that we must even have recourse to it in order to uphold the true (eucharistic) presence which I admit I am not able to reconcile sufficiently well with the view that puts the essence of bodies in a completely empty extension. 'It was by this sole light that in the autumn of 1691 that he prided himself "on having established a new science" which he called '*la Dynamique.*'[4]

Undoubtedly, that is also the reason why the *Essay* of 1692 remains no more than a first attempt in this new science.

1. Manuscripts at Hanover BR Foucher, fol. 23 (*Die phil. Schriften von G. W. Leibniz*, ed. Gerhardt, I, p. 403); *Journal des Sçavans,* 2 June 1692.
2. M. Guéroult, *Dynamique et Métaphysique Leibniziennes,* Paris, 1934, p. 97.
3. Manuscripts at Hanover HS Theology I, XIX, 7, fol. 527. Foucher de Careil, *Œuvres,* II, p. 45.
4. Letter to Pelisson, September–October 1691. See above, p. 64, note 1.

III

The General Rule
for the Compounding of Motions

Account of the text

Comparisons and dating

The second copy made by Des Billettes, which we discovered in the archives of the *Académie*, has the title *Règle générale de la Composition des mouvements par M. d. L.*, and comprises a text similar to that published in the *Journal des Sçavans* for 7 and 14 September 1693. Both texts, that of Des Billettes and that in the *Journal des Sçavans*, are reproduced in the Appendix, being arranged synoptically so that they may be easily compared. The text of Des Billettes is given in its entirety. The blank spaces have been arranged so as to set the textual variations between two versions immediately opposite each other; those portions that are identical with the earlier text have not been reproduced.

A quick inspection is sufficient to show that the text of Des Billettes served as the basis and that the variations of 1693 are not so much corrections as complementary explanations or glosses. For that reason, it is absolutely certain that the text of Des Billettes is the earlier one.

A comparison with the commentaries at the end of problem 1 makes it possible to be even more definite. In the text of Des Billettes, Leibniz quotes Fatio de Duillier and Huygens in connection with Tschirnhaus. The text of 1693 does not mention Huygens, but adds the Marquis de l'Hôpital to the other two names. 'Finally, the Marquis de l'Hôpital has provided the most general statement that could be desired on this subject, based on the new method of the calculus of differences.'

It was through the intermediary of Malebranche (letter of 8 December 1692) that the Marquis de l'Hôpital came to have connection with Leibniz. On 28 April 1693[1] the latter wrote to his new correspondent

---

1. Gerhardt, *Leibn. math. Schr.*, II, p. 237.

saying that he had found a simple solution to the problem of tangents to curves defined by a multipolar equation. De l'Hôpital replied on 15 June 1683[1] by sending the solution to which Leibniz refers in his article in the *Journal des Sçavans* for September.

That article had been written taking into account the communication from De l'Hôpital and had been composed very promptly, for the lapse of time between receipt of the communication and despatching the manuscript for the *Journal des Sçavans* must have been very short. Leibniz must, then, have had in hand an original copy which simply needed revision, completion, and editing in order to remove whatever might seem badly expressed or inadequate.

Indeed, the outline of the copy of Des Billettes was already in existence, with exactly the same expressions, in a letter from Leibniz to Huygens dated 13 October 1690.[2] Hence, it seems almost certain that the text in question was written between 1691 and 1692.

However, we may add, to the prospect opened by the present study in chapter I, several most interesting facts furnished by Leibniz himself. In fact, he wrote to Bossuet in 1693[3] as follows: 'The late M. Pelisson, having much appreciated what I had treated of in my dynamics, engaged me to send him a specimen for communication to the gentlemen of the *Académie Royale des Sciences* in order to learn their opinion: but he could not obtain it, although M. l'abbé Bignon and the late M. Thevenot were concerned in the matter. For that reason M. Pelisson consented that I should appeal to the public by putting a rule for the compounding of motions in the *Journal des Sçavans*.' He wrote also to the Marquis de l'Hôpital; the letter is undated, but can be assigned only to the autumn of 1693[4]: 'Several months ago I sent M. Pelisson my general rule for the compounding of motions from which I had deduced my rule of tangents by foci, with the design of having it inserted in the *Journal des Sçavans*; but as his death occurred unexpectedly, I have since sent it afresh.' De l'Hôpital had no need of that announcement seeing that on 21 September 1693,[5]

1. Gerhardt, *Leibn. math. Schr.*, II, p. 243.
2. Gerhardt, *Der Briefwechsel von G. W. Leibniz . . .*, Berlin, 1899, I, p. 603. Huygens, *Œuvres complètes*, IX, p. 519.
3. Letter from Leibniz to Bossuet [n.d.], 1693. A. Foucher de Careil, *Œuvres*, I, pp. 437–446.
4. Letter from Leibniz to De l'Hôpital [n.d.]. Gerhardt, *Leibn math. Schr.*, II, p. 248.
5. Letter from De l'Hôpital to Johann Bernoulli, 21 September 1693. *Der Briefwechsel von J. Bernoulli*, O. Spiess, Basel, 1955, pp. 188–190.

when informing Bernoulli of his own solution of the tangent problem, he mentioned the article of Leibniz in the *Journal des Sçavans*. He also expressed his hope of seeing 'his' theorem published in the memoirs of the *Académie*. So Leibniz was quite right to make haste for his own publication, otherwise De l'Hôpital would have been beforehand. The scope of his *Règle générale de la Composition des mouvements*, nevertheless, went far beyond the particular problem of the construction of tangents to certain curves, which we shall have to consider in detail; but it is the purpose of that work which must be specially referred to at the outset.

Indeed, on his own admission, it was after having ascertained his setback at the *Académie* with the *Essay de Dynamique* that Leibniz had the idea of 'appealing to the public' with an article in the *Journal des Sçavans* on the compounding of motions. According to the evidence of a letter to Foucher in May 1687,[1] that question had been linked for a long time with his own investigations towards a comprehensive doctrine of dynamics. Consequently, he certainly had amongst his papers whatever was necessary to make an attack on the resistance of the French circle with a 'specimen' of his ideas on mechanics, which would be less difficult at first sight and not so extensive as the *Essay de Dynamique*. It was simpler to secure publication on that subject and furthermore served as a feeler. Having obtained Pelisson's agreement, he then sent him a text which is without doubt the one that is preserved in the copy made by Des Billettes. The death of Pelisson on 7 February 1693 prevented publication, and it was with the double purpose of bringing the delayed project to a conclusion and of forestalling De l'Hôpital's initiative that Leibniz definitely hastened at the beginning of July 1693 to complete his second written account.

Hence, the text copied by Des Billettes can be definitely assigned to the end of 1692, and it must be remembered that the text in question, though it relates to a matter of kinematics, was integrated both with the general dynamics of Leibniz and his attempt to penetrate the French intellectual circle.

In order to judge what it contributes to a knowledge of his thought, we must first of all place it in relation to the texts to which Leibniz refers, that is to say, we must examine the texts and forget dynamics

---

1. Letter from Leibniz to Foucher, May 1687. A. Foucher de Careil, *Lettres et opuscules inédits de Leibniz*, 1854, pp. 69 *et seq.*

for a while in order to participate in the discussion opened by Tschirnhaus on a problem concerning the construction of tangents.

The problem of Tschirnhaus

Ehrenfried Walther von Tschirnhaus,[1] famed for his work on caustics and the construction of optical glasses, had been a foreign associate of the *Académie Royale des Sciences* at Paris since 1682. At the beginning of 1687 he published at Amsterdam his work entitled *Medicina mentis seu tentamen genuinae logicae in qua disseritur detegendi incognitas veritates*, which was dedicated to Louis XIV, and which reveals a close study of Descartes and Spinoza. The first part of the work is rather short and deals with the four principles of philosophy and knowledge; it then continues with a rather long study of fundamental truths and the main problems in geometry, which is clearly intended to be a teaching manual, a work of practical logic for the use of examinees in geometry. It is in this second part that we find the problem of constructing tangents to defined curves by means of 'foci'.[2]

The figures in the text make it possible to complete what is lacking through vagueness in the definitions as stated. Tschirnhaus imagines a thread of constant length whose ends are fixed at given points and which is kept in tension by a stylet. It is required to investigate the tangent to the geometric curve described by the point of the stylet. If the two points of attachment coincide (Fig. 1), it is obvious that we have a circle whose tangent, perpendicular to the radius, makes equal

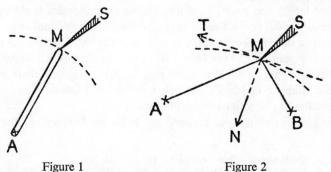

Figure 1   Figure 2

1. Ehrenfried Walther von Tschirnhaus, born 13 April 1651 in Haute-Lusace, died 11 October 1708 in Saxony.
2. See Huygens, *Œuvres complètes*, IX, p. 159.

angles on each side of the latter. If the two points of attachment are separated (Fig. 2), we have an ellipse whose tangent bisects the angle of the *radii vectores*, and again makes equal angles with them. However, with two separate points of attachment, matters can be complicated by causing the thread to pass round a third fixed point (Fig. 3), and we then obtain a new curve whose equation is defined by the following relation between the *radii vectores*

$$r_1 + 2r_2 + r_3 = l.$$

Tschirnhaus gives his solution for constructing the normal MN to that curve. It depends on drawing a circle with centre M which cuts

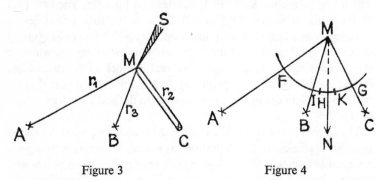

Figure 3   Figure 4

the *radii vectores* in F, I, G (Fig. 4), then finding the mid-points H and K of the circular arcs FG and IG, and finally the mid-point of the arc HK. According to Tschirnhaus the normal MN passes through the last mentioned point. Notwithstanding the statement that there is a proof, it will be fruitless to seek anything worthy of that name in the text of the *Medicina mentis*, as well as in the letter sent by the author to Huygens on 12 May 1687.[1] What is quite clear is that Tschirnhaus believed that he could proceed step by step, progressively increasing the number of 'foci', through a series of 'bisections' of angles, finalized in bisections of circular arcs.

The objections put forward by Fatio de Duillier[2] in the April

1. See Huygens, *Œuvres complètes*, IX, pp. 134–144.
2. Nicolas Fatio de Duillier, born at Bâle in 1664, citizen of Geneva, went to Paris at the age of 18 in order to study astronomy under J. D. Cassini; he settled in London in 1687 and became a Fellow of the Royal Society in 1688. He is noted for the part he played in the dispute between Leibniz and Newton. He died in Worcestershire in 1754.

number of the *Bibliothèque Universelle et Historique* for 1687[1] are therefore fully justified. Fatio said: 'He considered his method to be a good one, for the want of examining sufficiently closely an idea *that seems true by induction* and which in practice seems to be not far from the truth, but nevertheless does not agree at all with geometrical accuracy, except in some special cases and when the lines described by the threads are of the more common kinds.'

Fatio was not the only one to make such a criticism. Leibniz wrote to Tschirnhaus[2] to tell him that his method 'could be successful but rarely and that his enumeration of the curves of each degree is not satisfactory'. In order to clarify the account we shall consider the intervention of Leibniz in the discussion separately, and first of all deal with Fatio de Duillier and with Huygens.

Fatio de Duillier's solution

In the article mentioned above (April 1687) Fatio de Duillier first of all proposes to construct the tangent to the curve defined in bipolar co-ordinates by the equation $\lambda r_1 + \mu r_2 = $ Const, λ being the number of strands of thread on segment AC and μ the number of strands of thread on segment CD. First of all, he states the result: on segments CA and CD mark off two equal lengths CM and CP, then find the point N which divides MP in the ratio of μ to λ. CN is the normal to the curve described by C (Fig. 5).

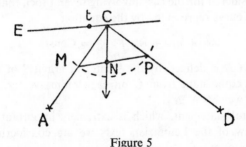

Figure 5

The proof, which follows immediately, consists in taking any point t, except C, on CE perpendicular to CN, and showing by geometrical

1. *Bibliothèque universelle et historique*, April 1687, V, pp. 25 *et seq.*
2. Letter from Leibniz to Huygens, 13 October 1690. See p. 67, note 2.

arguments (evaluating distances and similar triangles) that the point
t is 'outside the curve'. The conclusion is that 'no point of the curve,
except *C*, is on the straight line *CE*' and that *CE* is definitely the
required tangent.

Fatio compares his solution with that of Tschirnhaus. The bisections prescribed by the latter amount, quite correctly, to dividing the
circular arc *MP* in the ratio required by Fatio (namely, μ to λ),
whereas it is the segment of the straight line *MP* which must be
divided. That is the whole difference; but it is important because the
arc should not be confused with the chord. Naturally, with reference
to the direction of the normal thus determined, there can be agreement between the two solutions for special cases, and in general the
solution given by Tschirnhaus 'is not far from the truth in practice'.

Nevertheless, that solution is quite wrong. 'Indeed,' said Fatio, 'it
can be proved that tangents to all geometrical lines can be found by
solving an equality where the unknown has only a single dimension',
that is to say, by solving a linear equation. If the solution given by
Tschirnhaus was correct, 'it would have proved that the section of a
circular arc in a given ratio can be done with a ruler and compass'.
No doubt, the argument would need to be put in a more suitable
form in order to have its full force, but its merit is undeniable; so we
thought it desirable to reproduce it, for it shows the capability of the
kind of opponent that Tschirnhaus had met in the person of Fatio de
Duillier.

The article written by the latter ends with a generalization of the
solution considered for the case involving several foci, that is to say,
the case of a curve represented by the equation

$$\lambda r_1 + \mu r_2 + \nu r_3 + \cdots = \text{Const.}$$

The normal *CN* is defined by the 'centre of gravity' of the points
M, *P*, *Q*, ... equidistant from *C* on each radius vector and acted
upon by the weights λ, μ, ν.

This general statement, which is extremely important from the
point of view of the Leibnizian texts we are considering, is not
followed by any proof.

For the proof, we have to wait for Fatio's second article which
appeared in the April number of the *Bibliothèque Universelle et
Historique* for 1689.[1] That article replies to Tschirnhaus, whose de-

---

1. *Bibliothèque universelle et historique*, April 1689, XIII, pp. 46–76.

fence merely consisted in saying that the curves with two foci considered by Fatio are at most of the fourth degree and, consequently, that his method is not applicable either to an 'infinity of lines'. Fatio readily agreed; he wanted 'to prove that a certain method', put forward as being general, was not correct and it was 'by deliberate intent' that he sought 'easy examples of restricted application' where that rule shows itself to be defective. 'If M. de Tschirnhaus has not realized that his method involves dividing a given arc in a given ratio, it is because he has not had time to give sufficient consideration to his doctrine and to realize the consequences.' 'To do him justice' as he seemed to expect, one can only say 'that he had a very fine and far-reaching idea of a theorem that was not yet known to him', and that he provided 'the occasion for its discovery'.

The proof of that theorem is a fine example of the method of infinitesimals, and we cannot but admire its restrained elegance.

Let m be the point of the curve defined by the equation $\sum \lambda \, . \, ma =$ Const, having any number a of required foci. It is required to find the tangent to the curve at the point m. If t be a point 'infinitely close' to m on the tangent, 't may be regarded as a point on the curve itself'. On going from m to t, the algebraic sum of the lengthenings and shortenings of the *radii vectores*, qualified by the coefficients $\lambda, \mu, \ldots$, is therefore zero. Now, the perpendiculars ti on the *radii vectores* can be compared with the circular arcs having the foci as centre, for

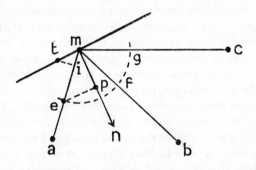

Figure 6

example a. Thus, mi represents the change in the radius vector on going from m to t, and we have $\sum \lambda \, . \, mi = 0$. If we then consider the point e on ma at a given fixed distance from m, and the perpendicular ep to the normal mn, the similarity of the triangles (mep) and (tmi)

enables us to conclude that $\sum \lambda . ep = 0$. Therefore, mn is definitely 'the axis of gravity of the points $e, f \ldots$ acted upon by the weights $\lambda, \mu, \ldots$'.

Fatio generalizes the case of curves defined by $\sum \lambda . r^p = \text{Const}$, and states that the weights acting on the points $e, f \ldots$ are $\lambda r^{p-1}, \ldots$ etc., even when p is fractional He does not prove it, undoubtedly because he thought it unnecessary after the preceding explanation. It is quite obvious, in fact, that the differential equation should be replaced here by: $\sum [\lambda r^{p-1}] . mi = 0$.

Fatio and Huygens

Had Fatio really conceived things in that way as early as 1687? We have here the instructive story of a curious reluctance on the part of the author of so fine a solution. We read in the papers of Huygens:[1] 'On the 12th or 13th March (1687) M. de Duillier communicated to me his method of tangents for the curves of M. de Tschirnhaus. The following day I showed him my correct proof. On Sunday the 16th I found that the perpendicular to the tangent must pass through the centre of gravity of all the threads serving to describe the curve allotting equal portions to them from the given point and proved it in the case of 2 or 3 threads. On Monday the 17th I told M. de Duillier about it, who was at first minded to deny the matter, he having been very near however to finding the same thing, but having afterwards rejected it.'

In his article of 1689, Fatio de Duillier admits that at the time when he discovered his theorem 'a famous mathematician from Holland was on the way to discovering it'. 'Indeed, he had proved it for a small number of foci and he understood how he would be able to do it in stages for more composite lines. He was using the same principle that I do for my proof and which I had communicated to him. When he was occupied with his own proof, it happened, I do not know how, on account of the disorder of the papers on which I had made my investigation, that I started to have doubts about that theorem. He, having told me that he found it to be true, I straightway recognized the fact on casting my eyes on the calculations that I had made and I realized that I had not had any reasonable ground to doubt it.'

The absence of proof, which we have stressed, in the article of 1687, is therefore explained in the light of the quotations which we

1. Huygens, *Œuvres complètes*, IX, p. 181, item No. 2469.

have brought together. In March 1687, Fatio was still uncertain of the accuracy of an argument the essential elements of which he nevertheless possessed and which had a geometrical basis. It was the confirmation provided by Huygens that gave him confidence, and the explanations he offered could be regarded as rather curious: that disorder of his papers which would be the cause of his first doubts, the returned confidence 'on casting my eyes on the calculations that I had made'. It is best not to take them literally, but to remember that at the period when geometry and the infinitesimal calculus were still in their infancy, it was nothing out of the ordinary for an excellent mathematician to express doubts on the validity of his arguments until fortified by the positive agreement of a very great master.

This positive agreement relates to the fact that the normal to the curve passes through the centre of gravity of the points defined as stated and acted upon by suitable weights. According to Fatio, it relates also to the principle of the proof; but is this latter point true, and what part did Huygens play in the discovery? We must now make a careful examination of the texts which are available in the complete works of Huygens in order to try and answer those two important questions.

We have quoted above the most immediate facts from the notes of Huygens in March 1687, which notes assume the presence of Fatio at The Hague and allow us to follow the development of the discussion day by day. Those notes reveal even much more. Having stressed the uncertainty of Fatio on the 17th, Huygens adds: 'Nevertheless, what he had found concerning the sum of the sines served to give an easy proof of the theorem of the abovesaid centre of gravity and it was very fine. See the previous page.' What we find on the preceding page is merely a draft of the proof of 1689. Only, instead of replacing the infinitesimal changes *dr* in the *radii vectores* by the proportional distances *ep*, Fatio then replaced them by the sines of the angles of the *radii vectores* with the normal. 'He had found', continues Huygens, 'the centre of gravity of all the points (Fig. 7). Then he will regard the sum of all the perpendiculars drawn from a point on the line *AB*, if it were perpendicular to the tangent, as being equal on both sides of that line. Then, he thought that those distances from the centres of gravity of the threads to the point *B* being equal on both sides, the fact did not agree with the centre of gravity. However, if he had drawn the sines on to *AB* from the points *D*, he would have seen that they were each equal to the perpendiculars from *B* on to the lines

AN, and therefore that *BA* was the true axis of gravity of the threads.'

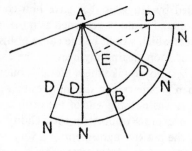

Figure 7

Fatio had therefore sensed the result and sought to justify its correctness *a posteriori.* It is convenient to recall here his first proof limited to the simple case of two foci. The result, namely, that the normal to the curve passes through the point that divides the segment *MP* in the inverse ratio of the numbers λ, μ, was called by him a generalization introducing the centre of gravity of points such as *M, P,* here designated *N.* Hence, there is no need for surprise at the course that was followed. Fatio considered the centre of gravity of the points *N* and sought to prove that that point always occurs on the normal to the curve.

By replacing the sines of the angles by the distances to the *radii vectores* of a point *B* on the normal in the equation deduced from $\sum dr = 0$, Fatio carried out a perfectly correct calculation, for the distances in question are indeed proportional to the sines. It was then necessary for him to establish that the centre of gravity of the points *N,* or of the threads (the ambiguity of this word 'or' being of no importance on account of the similarity of the centre *A*) complies with such an equation, that is to say, that the sum of the distances from that point to the threads is definitely the same 'on both sides of the line' which joins it to *A.* Now, the centre of gravity in question is, by definition, such that the sum of those distances to the points *N* (or to the centres of gravity of the threads) is the same 'on both sides' (equation for moments). The doubt that was raised in his mind is therefore easily explained. It did not seem that the same equality must be realized both for distances perpendicular from the point under consideration to the threads and for oblique distances which were the last ones to be considered. In both cases it was unquestionably a matter of the summation of distances, but Fatio's meticulousness seems justified. He did not notice what Huygens immediately pointed out, namely, that in the equation of sines these may be replaced by the distances *DE*, which immediately shows that the required normal *AB* is a straight line whose moment is zero with respect to the threads *AN* and therefore passes through the centre of gravity of the points *N*.

In short, Fatio's uncertainty resulted from a defective way of replacing the sines in his equation by proportional lengths. His first choice between two equivalent interpretations was not the right one, namely, that which immediately reveals the result, and that which he himself gave later in his article of 1689. It was Huygens who found the right interpretation.

The solution of Huygens

It might be thought that the questions we raised earlier on have been adequately answered. However, we can still learn something from a study of the ideas of Huygens.

On a sheet dated 12 November 1687,[1] Huygens wrote: '*Si trahantur omnia fila aequalia ab aequalibus ponderibus, sitque A centrum gravitatis punctorum omnium extremorum seu linearum ipsarum aequalium, manebit nodus seu punctum A ex nostro theoremate. Hinc probari potest summam istam filorum aequalium esse minimam, quia alias pondera trahentia possent descendere mutato loco A nodi; et ideo descenderent* (Fig. 8).

Quod si ita manent, manebunt etiam licet aliqua fila producantur, ut AB in C. Ergo et linea AC eum reliquis est summae brevissimae.

Ergo quod in plane demonstratur ex problemate Tangentium Fatii et nostro, hic etiam in solido verum esse evincitur.'

The starting point for the consideration of this text is therefore the equilibrium figure formed by equal weights pulling on threads of equal length tied together at the centre of gravity of the points of attachment or of the small pulleys which must be assumed for transmitting the action of the weights. Huygens deduces that the sum of the lengths of the threads in this equilibrium position is a minimum, because, if it were otherwise, then, on displacement of the point *A*,

1. Huygens, *Œuvres complètes*, IX, p. 183.

the weight would fall downwards and as a result equilibrium would not occur in the position indicated, there would be in fact a movement upsetting the equilibrium. Some few words are needed in order to explain more fully this very concise reasoning. If the sum of the

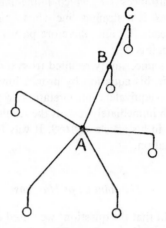

Figure 8

lengths of the threads were not a minimum, then, on displacing A, there would be no compensation between the lengthening and shortening of the various threads, and the shortenings taken all together would prevail over the lengthenings. The corresponding falls of the weights would prevail over the upward movements, the centre of gravity of the whole collection of weights would fall. There we have the indication that the starting position regarded as the equilibrium position would be inconsistent. The correctness of this reasoning is not obviously reached if, instead of attaching the thread AB at B, we place the attachment or small pulley for transmitting the weight at any point C on the straight line AB. Such a modification is without effect on the equilibrium conditions of the first figure considered and it alters nothing in the lengthening or shortening of the corresponding thread for a very small displacement of the point A. We can then dispense with the initial condition where the threads are of equal length. The lengths may be anything. The position under consideration is in equilibrium if the threads are tied together at one point A which is the centre of gravity of the points B located on the threads at any same given distance from A. That condition being satisfied, the sum of the lengths of the threads is a minimum for the position under consideration. That being so, if we direct our attention to one of the n threads constituting the arrangement, we find that a small displacement of A normal to that thread does not alter its length and is then related to the other $(n-1)$ threads by an exact compensation in lengthening or shortening, or by an equation of the form $\sum dr = 0$, where r represents the lengths of the $(n-1)$ threads or the distances from A to the points of attachment. The isolated thread obviously passes through the centre of gravity of the $(n-1)$ points B corresponding to the other $(n-1)$ threads, seeing that A is the centre of gravity of n points B; and on the other hand, seeing that $\sum dr = 0$ for the intended displacement of A, that displacement, which is normal to the isolated thread, corresponds at the same time to a geometrical definition with respect to the other $(n-1)$ threads, namely, $\sum r = $ Const. Thus, the problem of constructing the normal to the geometrical position of the points defined by an equation of the form $\sum r = $ Const with respect to as many foci as required is solved *whether or not the radii vectores r be in the same plane.*

That was the final state of the solution given by Huygens seven months after the discussion with Fatio; we cannot but admire the elegance and power of the proof. However, the instructive exposition does not restore the order of the discovery, precisely because it proceeds by deduction from a problem in mechanics far removed *a priori* from the purely geometrical question to be investigated.

To understand the matter more clearly, we must go back to the elaboration of that theorem of equilibrium which Huygens describes as his own: '*ex nostro theoremate*'.

That theorem is to be found in a memoir of the *Académie des Sciences* for 1667, and published in 1693 by La Hire in a collection entitled *Divers ouvrages de Mathématiques et de Physique par MM. de l'Académie Royale des Sciences*; the title of the memoir is '*De potentiis fila funesve trahentibus*'. The enunciation is as follows: '*Datis positione punctis quotlibet sive in eodem plano fuerint sive non, si a puncto, quod eorum commune est gravitatis centrum, ad unumquodque datorum fila extenantur eaque singula trahantur a potentiis qual sunt inter se ut filorum longitudines, fiet aequilibrium manente nodo communi in dicto gravitatis centro.*'

It is necessary to consider the proof in detail. Let A, B, C, D, E be the given points, which may or may not be in the same plane. Huygens allots the same 'weight' to them and determines their centre of gravity step by step: F the mid-point of AB, G such that

$GC = 2GF$, H such that $HD = 3HG$, etc. In that way he carries out the *geometrical construction* for the centre of gravity K of the whole set of points. He then says, if we 'extend' the threads from the point

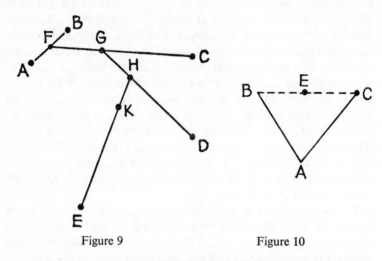

Figure 9   Figure 10

K as far as A, B, C, D, E, the threads being 'pulled by forces proportional to their lengths', the whole set of threads and the forces under consideration will be in equilibrium. Once more, the reasoning is carried out step by step. The forces AK and BK are equivalent to $2FK$, $2FK$ and CK are equivalent to $3GK$, etc., and the forces corresponding to the points A, B, C, D are finally equivalent to $5HK$. This last 'is directly opposed' to the force EK, and equilibrium of the whole arrangement immediately follows (Fig. 9). Huygens refers explicitly to a proposition placed at the beginning of the memoir: if two threads AB and AC exert on A pulls proportional to the length of the threads multiplied by the numbers N and O, the two pulls taken together are equivalent to a pull of $AE \times (N + O)$, E being the point that divides BC in the 'reciprocal' ratio of N to O (Fig. 10).

The sequence of the propositions is therefore perfectly logical and easy to follow; but we must go and find the starting point in that part of the memoir of September 1667 which was not published in 1693. The starting point is the following very brief text, accompanied by a suggestive diagram[1] (Fig. 11).

1. Huygens, *Œuvres complètes*, XIX, p. 51, item No. VII.

'*Sit nodus in C. Ergo PE raccourci de DE, QE de AE, SE allongé de BE. Oportet DE in p + AE in q − BE in s ∝ 0.*'

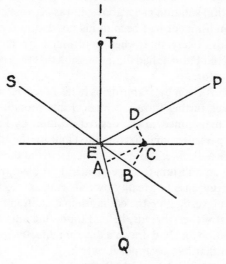

Figure 11

The figure admits of no doubt. The set of forces *p, q, s* directed along *PE, QE, SE* respectively being in equilibrium is regarded as equivalent to a weight suspended by a thread of given length at the vertical of the point of suspension *T*. The single small displacement compatible with the arrangement and without effort is an horizontal displacement *EC*. As a result we have two shortenings *DE, AE* for the threads *PE* and *QE*, and a lengthening *BE* for the thread *SE*. There must be compensation for the corresponding efforts and Huygens writes without hesitation an equation for virtual work: the sum of the products of the forces by the differences of the *radii vectores* must be zero.

From that he deduces the equilibrium condition for three concurrent 'forces', then the proposition giving the force equivalent to two given forces, using arguments to which we shall revert shortly, but which will not be more fully described at this point in order not to confuse an already complex subject.

The essential thing to be noted now for the purpose with which we are concerned has been obtained. The starting point for Huygens in

1667 was a problem of the equilibrium of three stretched threads, and that problem was solved thanks to considerations which were astonishingly close to those required for the solving of Fatio's problem in 1687. If Huygens were able to raise Fatio's doubt, and then to work out the beautiful solution given above, it was because all the elements for its accomplishment had been in his possession for a long while; and it is most likely that when confronted by the diagrams of Tschirnhaus and Fatio he had the impression that they were something he had seen before.

The diagrams given by Tschirnhaus in his *Medicina mentis* possibly led him to seek further for a mechanical interpretation. In fact, we find a cord maintained in the desired position by links (points of attachment, passage round fixed points) and by the stylet whose tip must describe the curve. It is not a great step from there to substitute a problem in equilibrium for a geometrical problem, especially for a master like Huygens; but reality exceeds conjecture. In 1667, when Huygens was investigating the equilibrium condition for three forces or for three stretched threads, he had found the answer by the least possible equation and had drawn a diagram directly applicable to the new problem that had been put forward.

That 'same principle' which Fatio claimed to have been communicated to him and which, as we have seen, is equivalent to $\sum dr = 0$ for a displacement of the stylet over the whole length of the curve, was quite clearly of a type with which the mathematician from Holland was well acquainted. Furthermore, seeing that he had used similar reasoning to obtain theorems expressing the equivalence of concurrent forces and the equilibrium of stretched threads by introducing the concept of centre of gravity, we can quite understand that he had no difficulty, in 1687, in expressing the result for constructing the normal to a curve with the help of a centre of gravity. He was obviously moving in a field where the linking of geometrical and mechanical propositions had no secret for him.

However, it should be noted that when Huygens introduced the centre of gravity in order to express the force equivalent to two given forces, that was only a result obtained by geometrical means. The transformation for the equilibrium condition of three forces shows that the force equivalent to two given forces must pass through the point which divides a segment in a simple ratio. It is that geometrical property which proclaims a centre of gravity in the point in question. Similarly, and we have stressed the matter, the proof of the theorem for the equilibrium of threads attached to a point *A*, which is the centre of gravity of points located on the threads at a like distance from *A*, introduces no mechanical property of the centre of gravity. We see, then, how much care is needed if we want to distinguish between the positions of Fatio and Huygens at the time of their discussion in March 1687.

It is true to say that Huygens differs from Fatio, as we have already said, by considerable acquaintance with a field of knowledge in which geometrical and mechanical propositions are linked together. It is true to say that the concept of equilibrium and its interpretation in various ways provided him with the clue; but when he expresses the results by means of centres of gravity, he does so, because certain points correspond to the geometrical definition of such centres; and in that respect he differs from Fatio.

Whereas in 1689 Fatio still only knew how to deduce the result of his theorem by reasoning of a geometrical character, by proving that the normal to the given curve is a straight line whose moment is zero, Huygens, on the other hand, as early as 1687, had worked out a more avowedly mechanical solution. If the reader will refer to that noteworthy text and to our commentary, he will immediately note this essential difference: the final arguments that Fatio discovered in considerations of infinitesimals in geometry, Huygens finally obtained from the fact that the height of the centre of gravity of a system of points in equilibrium becomes a minimum.

The solution of Leibniz

The ground has now been cleared for our consideration of the part played by Leibniz in this matter. A short passage from the letter written to Huygens on 13 October 1690 has been quoted above; it discloses that Leibniz agrees with Fatio in stating that the rule given by Tschirnhaus applies only occasionally. We must now look at the matter more closely.

'I set about trying to find a better rule for determining tangents by foci and threads,' Leibniz continued, 'and I found it. However as regards publication, I have been forestalled by M. Fatio Duillier, on account of which I am not greatly vexed, for it seems to me that he has much credit. Nevertheless, I shall tell you my way of proceeding.'

We shall consider that way of proceeding in more detail. For the moment, let us note that Leibniz states that he had sent comments to

Tschirnhaus immediately 'after publication of his work' and at the same time that he had found a better rule, which rule he had not been able to make public on account of Fatio's article. That statement agrees with the commentaries to problem 1 of the text under consideration.

It ought to be possible to check the truth of the statement made by Leibniz from correspondence and the printed texts of the individuals involved. Unfortunately, the facts are not particularly clear.

Relations between Tschirnhaus and Leibniz were certainly friendly in 1686. Foucher wrote to Leibniz at the beginning of 1687:[1] 'I have had lent to me the book, *de medicina mentis et corporis*, by your friend Thirnous. So far, I have read only the beginning and I find it excellent'; to which Leibniz replied in May of the same year:[2] 'M. Tschirnhaus used to be a far better Cartesian than he is now. Still, I have contributed something to open his eyes.' With regard to the *de medicina mentis*, he added: 'There are many good ideas in the book by M. Tschirnhaus. His way of treating foci which may be lines or points is a very fine device, but there are some particularities and consequences where he takes too much for granted. For he believes he can easily determine the number of all curves of every degree, which I know cannot be so. I should have liked to have known of his design to have the work printed in order to have opened his eyes earlier. However, that does nothing to reduce the esteem in which I hold him.' However, that esteem did not stand the test of time, seeing that Leibniz wrote to Huygens on 13 October 1690 as follows: 'I, no more than you, Monsieur, have reason to be pleased with M. de Tschirnhaus, for it has happened to me more than once that he has forgotten that he has seen specimens of my work which he has given somewhere afterwards.' That Leibniz should agree with Huygens in accusing Tschirnhaus of a certain lack of constraint in using ideas which had been communicated to him obviously constitutes a disparaging judgement on the last named. That judgement would not exist were there not some estrangement between Leibniz and Tschirnhaus. The very titles of the memoirs presented by Tschirnhaus to the *Académie* between 23 December 1701 and 10 January 1702 are significant: '*Méthode pour trouver les rayons des développées, les tangentes, les quadratures et les rectifications de plusieurs courbes sans y supposer aucune grandeur infiniment petite*'

1. A. Foucher de Careil, *Lettres et opuscules*, 1854, p. 79.
2. *Ibid.*, p. 69 *et seq.*

and '*Méthode pour trouver les touchantes des courbes mécaniques sans supposer aucune grandeur infiniment petite.*' Tschirnhaus opens both these memoirs by a passionate controversy between Johann Bernoulli and the Marquis de l'Hôpital, in which controversy he reveals himself as an indomitable opponent of differential calculus. We can easily understand the endorsement written by Leibniz himself on the copy of an undated letter to Tschirnhaus, but which can be assigned to the period 1685–1687.[1] In that letter Leibniz praised the benefits of differential calculus. The endorsement in the margin reads: '*Ist nicht abgegangen*' (not sent). Leibniz undoubtedly indefinitely postponed sending the letter, because he considered that he was dealing with someone who was little inclined to accept new ideas and 'to have his eyes opened' to the extent that he first thought. The favourable opinion which he had originally of his fellow countryman was rapidly dissipated between 1687 and 1690.

Nevertheless, we should like to know to what precise point the criticism which Leibniz claims to have sent to Tschirnhaus in connection with the solution of the problem of tangents in *de medicina mentis* refers, besides the example which he claims also to have given in order to show the defective nature of that solution. Unfortunately, there is no trace in the surviving correspondence between Leibniz and Tschirnhaus; so we can only make conjectures.

In his reply dated 18 November 1690,[2] Huygens told Leibniz: 'I had part of M. Fatio's rule for the centres of gravity, as he himself admits in the journals. However, it was he who first pointed out to me the mistake made by M. de Tschirnhaus.' We have already seen above how Fatio was able, in fact, to detect that mistake with remarkable clarity. When we compare it with the passage from the letter of Leibniz to Foucher quoted above, we cannot but feel rather uneasy. All that Leibniz could say in support of his judgement on the solution given by Tschirnhaus is rather weak compared with Fatio's sharp criticism. Similarly with regard to 'those many good ideas' which Leibniz said he had found in the *de medicina mentis*. Politeness sometimes demands more than one means. However, there can be no

---

1. Gerhardt, *Leibn. math. Schr.*, IV, p. 507.
   Calculus differentialis ostendit non tantum quidquid ab aliis circa tangentes et quadraturas hactenus repertum est, sed et innumera, in quae nisi calculo meo usus (cujus nuper initia quaedam Lipsiam publicanda misi) non facile incidit quia isto calculo omnia mira brevitate et claritate oculis ac menti objiciuntur.
2. Huygens, *Œuvres complètes*, IX, p. 538.

illusion on what followed next. Leibniz straight away realized that the solution given by Tschirnhaus was radically wrong.

What excited his admiration in that solution was the 'fine device' which consisted in 'treating foci which may be lines or points', and he merely considered that the author 'takes too much for granted'. That vague and indefinite criticism is far removed from Fatio's vigorous expressions.

It is true that Fatio did not even concern himself with the generalization that so enchanted Leibniz, and that is why we have not spoken about it so far. When Tschirnhaus replaces a focal point by a closed curve, it must be understood that instead of the radius vector r we must consider the length l of a thread stretched from the moving point M on the curve in question to a fixed point on the focal curve (Fig. 12). For a curve with one single 'focus', defined by $l = $ Const,

Figure 12

that is to say for an involute of the 'focus', the tangent at M is normal to the straight part of the stretched thread, exactly as when the focus is a point. We have already seen that that was the basic property used by Tschirnhaus in order to elaborate the whole of his solution for the case of several foci, by means of a series of bisections. Consequently, we can easily understand why the generalization of his results to account for curves defined with respect to foci other than points, in the sense that has just been specified, did not call for any additional justification on his part.

In fact, that generalization, regarded as an exact argument, suffers from no objection. The differential dl for an infinitesimal displacement of the point M remains in effect the projection of that displacement on the stretched thread, and there is nothing that requires modification in Fatio's solution which is valid whether the foci are points or not. Was Fatio aware of that? It is impossible to say in view of the complete silence that he maintained with regard to the question of curved foci. We can only think that the matter was not beyond his capabilities, and that, with the principles of his solution, he was at least in possession of the means to provide the true reason for the accuracy of the generalization.

Obviously, that did not apply to Leibniz. Confronted with the solution of Tschirnhaus, his philosophic mind went straight to admiration of a result embodying more general definitions than those given by point foci, but he did not have a firm judgement on the fundamental, possibly because he had not given the matter sufficient attention. He merely had the impression that Tschirnhaus 'takes too much for granted' and that matters were more complex than rapid intuition would lead one to believe at first sight. The particular point on which he was ready, as a result of his previous investigations, to find Tschirnhaus at fault was not the one connected with the precise problem of determining tangents to the curves under consideration; it was those consequences by which Tschirnhaus believed he could 'easily determine the number of all curves of every degree'. Hence, the attention of Leibniz was not directly drawn to the profound validity of the arguments of Tschirnhaus.

What then must we think of the assertions in his letter to Huygens on 13 October 1690? Had he really, at the beginning of 1687, conceived a better solution of the problem of tangents? Had he really been forestalled by the publication of Fatio's article? We can only ask those questions without being able to give any clear answer. It is not impossible that after his letter to Foucher at the beginning of 1687, Leibniz had clarified his own views by discovering the example of which he speaks and where the solution given by Tschirnhaus was defective. It is not impossible that at the same time when Fatio was completing his solution for publication, Leibniz had found the same result by another way. Again, it is not impossible that on reading what Fatio had written, Leibniz was constrained to elaborate that new 'way of proceeding' to the solution which it is now appropriate to consider.

'I have found and proved this general principle', wrote Leibniz to Huygens on 13 October 1690, 'that any moving body having several directions at the same time must go in the line of the direction of the centre of gravity common to as many moving bodies as there are directions, if we imagine the single moving body multiplied as many times as are needed to make each direction function fully at the same time, and that the velocity of the moving body in the compounded direction must be to that of the centre of gravity of the [fictitious] device as the number of directions is to unity'. The application of that general principle to the construction of a tangent to the curves of Tschirnhaus is then given in terms exactly similar to those which are found in the text of 1692. 'The stylet which stretches the threads could be regarded as having as many directions equal in velocity amongst themselves as there are threads. For, in the same way that it pulls them, so it is pulled by them.'

We shall not hide the fact that those lines are difficult to understand on first reading. Huygens himself was not enraptured, seeing that he replied as follows on 18 November 1690: 'Your musing on tangents by foci appears to me most profound. Nevertheless, it assumes things that cannot be accepted as obvious. Although *such arguments may sometimes serve for discovery*, other means are subsequently needed for *more certain proofs*'.

It is possibly on account of that objection by Huygens, expressed in terms that must have gone home as far as Leibniz was concerned, that we owe fuller explanation of the general principle of the compounding of motions as contained in the texts reproduced here.

The most finished state of the work is provided by the text in the *Journal des Sçavans* for 1693. We have already remarked that that text was a consequence of direct contact with the Marquis de l'Hôpital. Without going into details, or offering anything that could serve that purpose, Leibniz announced to his new correspondent on 28 April 1693 that he had in hand a better and simpler solution than the one given by Tschirnhaus, a solution *based on a fine consideration of mechanics*. Warned of the difficulties of that *fine consideration* as a result of the remarks made by Huygens, Leibniz clearly sought to make his text more understandable when preparing it for publication.

We shall follow the version of 1693 which clearly explains the substance of the communication sent to Huygens in 1690.

A moving body *A* is supposed to be subjected to 'various tendencies' such that if each were acting individually, they would cause the body to travel with uniform motion in one second along the straight segments *AB*, *AC*, *AD*, *AE*, etc. In order to find the motion resulting from the simultaneous action of the different 'tendencies' Leibniz made use of a fictitious device. He imagines that the moving body is shared equally between the 'motions so as to satisfy them all together perfectly'. For example, if there are four 'tendencies', then 'each acquires only one fourth part of the moving body which must go four times further in order to make as much progress as if the whole moving body had satisfied each tendency'. The centre of gravity of the parts of the moving body progress also four times further and its displacement gives the required displacement of the undivided moving body A subjected to the action compounded of the different 'tendencies'. If G be the centre of gravity of the points B, C, D, E where the moving body A would converge if each tendency acted separately, $AM = 4 \times AG$ is the compound or resultant displacement.

In present-day terms, the general rule given by Leibniz for the compounding of motions is as follows. If $\overrightarrow{AB}$, $\overrightarrow{AC}$, $\overrightarrow{AD}$, $\overrightarrow{AE}$ be the velocities of the component motions to which the moving body A is subjected, then $\overrightarrow{AM} = 4 \cdot \overrightarrow{AG}$ is the velocity of the compound motion.

The proof follows immediately from the very definition of the centre of gravity G of the points B, C, D, E that $\overrightarrow{AB} + \overrightarrow{AC} + \overrightarrow{AD} + \overrightarrow{AE} = 4 \cdot \overrightarrow{AG}$. Nowadays, we could be tempted to do Leibniz the favour of not discussing his own proof for fear of getting lost in quibbles or insignificant details; but to be freed from that fear it suffices to note that in order to obtain such a simple result in pure kinematics, Leibniz felt himself obliged to employ dynamical considerations. The terms 'tendencies' and 'motions' are used indiscriminately and obviously involve the concept of mass. His proof, then, is worthy of closer study; at least, its nature is interesting and capable of throwing fresh light on his thought.

However, because a study of that feature would divert us for a while from our subject, namely, the solution of the problem of Tschirnhaus, and because it constitutes a quite separate matter in itself, we shall not revert to it until after having considered how Leibniz applied his general rule for the compounding of motions to the construction of tangents to curves defined in multipolar coordinates.

Leibniz said, 'We must consider that the stylet which stretches the threads could be regarded as having as many directions equal in velocity amongst themselves as there are threads, for in the same way that it pulls them, so it is pulled by them. Consequently, the compounded direction, which must be in the perpendicular to the curve, passes through the centre of gravity of as many points as there are threads. And those points, on account of the sameness of the tendencies are equidistant from the stylet and therefore fall at the intersections of the circle (with centre A and any radius) with the threads.'

In the first place, note the identity of that result with those of Fatio and of Huygens and then its accuracy. Leibniz, as did Fatio and Huygens, considered the points located on the threads at a given same distance from the point on the curve; he allotted to each of those points a weight equal to the number of threads superimposed in the same direction and took the centre of gravity of the whole arrangement. The normal to the curve passes through that centre of gravity.

In order to arrive at that result, Leibniz considers the point of the 'stylet' placed at A in equilibrium on the curve, the tangent to which it is required to construct, and subjected to equal forces of tension by all the strands of threads which converge there.

The resultant or compounded tension is normal to the curve, and it is consequently quite clear that Leibniz resolved the problem as one of equilibrium of a moving body constrained to remain on a curve. In order to find the direction of the compounded tension, he applies his rule for the compounding of motions, thereby confirming the confusion present in his mind between the dynamical and kinematical points of view. On the other hand, he assumes without further ado that the equality of action and reaction for each strand of thread results both from equality of the tensions of the threads amongst themselves and from equality of the imaginary velocities corresponding to those tensions. The doubt expressed by Huygens on the validity of such an argument was therefore fully justified, and it is necessary to note that if Leibniz directed his attention to clarifying his general rule for the compounding of motions, he left unchanged the method of applying it to the problem of tangents. In other words, he had not realized the difficulty of treating a problem in Statics and the compounding of forces by considerations relating to the compounding of motions and to kinematics.

To be sure, very little was needed to make the application satisfactory. From the equation defining the curve $\sum \lambda r = \text{Const}$, we have for an infinitesimal displacement of the moving point: $\sum \lambda dr = 0$. Now, the radial velocities are proportional to dr. They are therefore inversely proportional in magnitude to λ, the number of strands of threads to be considered in connection with each radius vector r. They are directed in such a manner as to oppose each other, some being positive, and others being negative, on their respective *radii vectores.* If we then consider velocities of the same magnitude, but all in the same sense (for example, all negative), it is obvious that the resultant velocity is perpendicular to the resultant velocity of the preceding case, namely, for an infinitesimal displacement of the moving point on the curve and also on the tangent. Consequently, the determination of the normal reduces to the determination of the velocity compounded of equal radial velocities, equal in number to the strands of threads to be considered in conjunction with each radius vector.

The solution of Leibniz could then be made correct by freeing it from the confusion of mechanics, as we have already stressed, by reverting to the differential equation for the curve. In that way, it could be furthermore generalized for the case envisaged by Fatio: $\sum \lambda r^p = $ Const. That generalization was nevertheless impossible in the form given by Leibniz with its confused, intuitive consideration of equal tensions along the threads.

However paradoxical it may seem in this problem concerning infinitesimal geometry, it is not Leibniz who is using 'his calculus'. It is Fatio and Huygens who rely appropriately on the differential equation of the curve under consideration; the former in order to make a direct geometrical study of it, the latter in order to give an explanation of it in accordance with mechanics. Leibniz allowed himself to be led astray by an alluring intuition and thought that he had got round the difficulty by means of a 'fine consideration of mechanics', which was not worth much.

In short, if he had really found that solution before having read what Fatio had written, he was right not to enter into dispute with him. There would have been nothing to gain thereby. He appears in a fresh light in this affair: that of a soulful individual, quick to submit to the lure of the flashes of brilliant intuition, but less equipped to carry the analysis of it to the required depths.

The final point on the subject of 'Fatio's theorem for tangents' is provided by the Marquis de l'Hôpital in his letter to Leibniz of 15 June 1693 and in his letter to Johann (I) Bernoulli of 21 September 1693. After all that we have said, some few words will suffice to characterize the conclusion of a solution which gave rise to controversy.

We have already mentioned that Leibniz in his letter to the Marquis de l'Hôpital of 28 April 1693 contented himself with reporting that he had found a better and simpler solution without giving him any indication that would put him in the way of finding it. Hence, the Marquis de l'Hôpital applied himself with his own resources to solving the problem; and he succeeded in providing the solution with a more general form which he charges Fatio with not having considered.

He says: 'Let *MPN* (Fig. 13) be a curve, such that having drawn the straight lines *PA, PB, PC* from any point *P* to the foci *A, B, C*, etc., the *relation* between them is *any relation*. If we take the differences of that equation, and having described a circle with centre *P* so that it cuts the lines *PA, PB, PC* in the points *E, F, G*, we regard

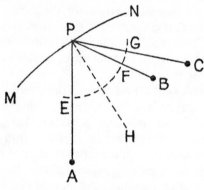

Figure 13

those points as being subjected to weights which have the same ratio amongst themselves as *the quantities that multiply the differences* of the lines on which they are located, then the line *PH* passing through their common centre of gravity *H* will be perpendicular to the curve.'

The principle of the proof is that of Fatio, but De l'Hôpital realized that the essential requirement for applying that principle lies in the differential equation of the curve. For that reason, being dissatisfied with Fatio's use of successive generalizations in order to ensure the result relating to more and more complex curves, as though such a procedure were justified in the nature of things, De l'Hôpital started immediately from the most general definition of the curve under consideration: $f(r_1, r_2, \ldots) = 0$.

In the differential equation for that curve, namely, $\sum \frac{\partial f}{\partial r_i} \cdot dr_i = 0$,

we have the general relationship which links the 'differences' of the *radii vectores* for an infinitesimal displacement on the curve, and the

92

interpretation given by Fatio remains valid in taking for the 'weights' acting on the E, F, G, etc., 'the quantities which multiply the differences', namely, $\frac{\partial f}{\partial r_i}$. Indeed, we may think that De l'Hôpital has little credit for having found that general expression. Fatio and Huygens, as we have pointed out, had known how to rely on the differential equation of the curve, and so had largely prepared the way for him. Nevertheless, the fact of the generalization on the part of De L'Hôpital provides evidence of his ability which is far from being negligible. His is the credit for the attention directed to that which made a success of the solution and also to that which formed its essential feature.

That Leibniz should have been outstripped on that matter in the use of 'his' own calculus of differences was unfortunate, but misfortune is not infrequently the fate of the greatest masters.

The compounding of motions

In order to have a better understanding of Leibniz, we must now revert to that general rule for the compounding of motions which involved him in an indifferent solution of the problem of tangents.

It is now necessary to say that the solution seemed to him to be a particular application of a very general principle, and to the extent that that principle was not sought for the requirements of the case, it is clear that any judgement must be modified.

In that connection, attention must be directed to two facts. In an article, *de Geometria recondita*,[1] which appeared in the *Acta Eruditorum* for June 1686, Leibniz proposed to accept amongst geometric curves those, such as the cycloid, which can be exactly described by a continuous motion. In referring to that article, Huygens wrote to Fatio on 11 July 1687[2] saying that he expected enlightenment on the method of tangents, 'a new discovery which will be fine if it be applicable to all geometric curves even if those of the type [described] by M. Leibniz are not covered by it'.

So, in 1686 Leibniz had given consideration to curves whose definition depended on mechanics and which according to the classifications of the period and the generally accepted views were not

1. *Acta Eruditorum*, June 1686, p. 292.
2. Huygens, *Œuvres complètes*, IX, p. 181.

regarded as being on the same plane as 'geometric' curves. He considered that there was no reason to establish a watertight division between the two categories of curves depending on the method of definition.[1] Huygens, on the other hand, seemed to have doubts, particularly as regards the treatment of the problem of tangents. Consequently, it is more than likely that the general rule for the compounding of motions on the part of Leibniz and its application to solving the problem of Tschirnhaus merged into a general preoccupation by the great German philosopher: namely, to prove that considerations of compounded motions make it possible to solve questions depending *a priori* only on geometry alone, and so to weaken the accepted distinctions between too well-defined categories.

The second fact which it is desirable to emphasize is the statement made by Leibniz in his letter to Foucher of May 1687, to which we have already referred.

'M. de Mariotte and several others have shown that the rules of M. Descartes on motion are quite remote from experience, but they have not shown the true reason. Furthermore, M. de Mariotte relies usually on the principles of experience, the reason for which I am able to show by my general axiom, on which, in my opinion, the whole of Mechanics depends.

'The Rules for the compounding of motions on which many rely in those matters are liable to more difficulties than one imagines.'

The general axiom, on which the whole of Mechanics depends in the opinion of Leibniz, was evidently the conservation of absolute force; and we can easily understand that the principles of Mariotte seemed to him to be based principally on experience when compared with that axiom. A grave fault in the eyes of a metaphysician! Nevertheless, Mariotte had the credit of suggesting an axiomatic form for Mechanics which prejudges nothing in regard to the co-ordination of force-motion. That was not so in the case of authors such as Bernard Lamy and Pierre Varignon.

The latter, in his *Projet d'une nouvelle Méchanique*, published in July 1687, took as a basis the compounding of forces by the parallelo-

---

1. It may be pointed out here that the curves considered by Tschirnhaus obviously resulted from a confused investigation of graphical methods of delineation which dispensed with ruler and compass. On the one hand, the connection with the mechanical aspect of the question is readily accounted for; on the other hand, preoccupation with evaluating those curves and their assimilation with 'geometric' curves is most understandable.

gram rule, deduced from the compounding of velocities. The dispute between Lamy and Varignon about priority on this matter is complicated and difficult, but for our present purpose is certainly not without interest. We are concerned with remembering what the dispute proves; namely, that in 1687, the idea of using the compounding of motions in order to discover therein a basic principle for studying forces was sufficiently in the air to take shape with various writers working independently.

Leibniz had encountered it, even if it were not so neatly and rationally expressed as in Varignon's work. So, it is not surprising that he was consequently led to reflect on the compounding of motions. The general rule which he pointed out in that connection and which we have explained above in accordance with the version of 1693 differs from the parallelogram rule only in form and by a more general statement. The parallelogram rule is strictly equivalent to the Leibnizian rule for the compounding of two velocities, and its repeated application obviously gives gradually, by recurrence, the result given by Leibniz for any number of component motions. Here again, it is not so much the result as the way in which it was obtained that it is advisable to study.

Leibniz considered that that way is 'liable to more difficulties than one imagines' for those who make use of the compounding of motions. Nevertheless, at first sight, it consists in a simple verification. If the moving body A be subjected to motions of velocity AB and AC (Fig. 14), everything takes place as if the moving body, when

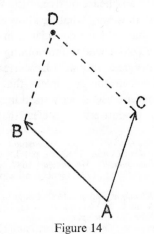

Figure 14

it has arrived at the point B after one second, had been subjected to a movement of uniform translation in the line AB from A to C depending on the velocity of AC. The moving body then arrives in actual fact, in the same time, at the point D, the fourth corner of the parallelogram constructed on AB and AC. Hence, the diagonal of the parallelogram is the compounded velocity AD.[1]

A simple verification at first sight, we said a moment ago. It is indeed so for anyone who is accustomed to separate ideas and to hold himself aloof from the solid mechanical conditions necessary for a realization of such a compounding of motions for the purpose of considering only the kinematic nature of the question. In a period when matters were not so, however, many difficulties failed to be removed. In particular, in order to accept the preceding argument, it was necessary to assume that two motions never interfere with each other during their simultaneous application.[2] Such reciprocal independence seemed to be certain when the directions are perpendicular to each other. It seemed doubtful when the directions make any angle with each other. On the other hand, for those who, like the Cartesians, adhered to the dogma of the conservation of the quantity of motion, in the scalar meaning of the word, such a result was disastrous, because the sum of the component quantities of motion would not give the quantity of compounded motion.[3]

In coming to the conclusion which we have emphasized above, Leibniz joined the Cartesians in their dislike of treating the compounding of motions as a problem independent of conditions realizable by mechanics, and hence by dynamics. We need have no surprise on that account, in view of what we know about his thoughts on mechanics, and of what we have considered in connection with the *Essay de Dynamique*. To one who found nothing so understandable as force and who, like his great opponent Descartes, tried to eliminate a consideration of time from the principles—that matter which in the manner of motion never exists strictly speaking seeing that its parts are never together as an entirety—it is clear that the problem of the

1. See Roberval, '*Observations sur la composition des mouvements,*' 1675. *Mémoires de l'Académie des Sciences*, 1730, VI, p. 90.

P. Varignon, *Projet d'une nouvelle Méchanique*, 1687, Lemma 3.

Nouvelle Méchanique, Posth, ed., 1725. General principle: Corollaries I, II, III, and Lemma, I, p. 13.

B. Lamy, *Traité de Mécanique*, 2nd ed., 1687. Letter to Dieulamant, July 1687.

2. See P. Varignon, *Nouvelle Méchanique*, Lemma, II, p. 14.

3. See P. Varignon, *Nouvelle Méchanique*, p. 24.

Nouvelles de la République des Lettres [Nuguet]. April 1705, Art. 2, p. 389 *et seq.*

compounding of motions acting simultaneously on a moving body cannot be anything but a problem in dynamics.[1] The problem must be solved, appropriately, in the perspective of the doctrinal unity created by the general axiom which its author flatters himself to have found for that science.

The moving body A subjected to various motions will consequently not be a geometric point for Leibniz, but a material point endowed with mass. The simultaneous application of the motions will be an action obliging that material point to 'satisfy' those 'various tendencies' all together, for it is definitely understood that there is no motion without 'tendency', the 'power to act', and which must constantly pass from one to the other of those two aspects of one single reality if we wish to understand the phenomenon. The total action to which the material point is subjected is a summation of the power to act of the various component 'tendencies'. We can therefore imagine the material point breaking up into as many equal portions as there are 'tendencies' to be satisfied. In order to fix his ideas, Leibniz assumed four of them. Each fraction of the initial moving body must then travel, he said, four times as far in consuming the force of each tendency as if that force were exerted on the entire moving body. That is to say, the principle that governs that part of the argument is the conservation of the quantity of motion in the vectorial sense. The centre of gravity of the parts is therefore four times further off from the initial position of the moving body than the centre of gravity of the positions reached by the entire moving body obeying each 'tendency' separately—that is merely a question of similarity. On the other hand, and here we have the crux of the proof, that centre of gravity of the parts is the position which the entire moving body reaches under the total action of the tendencies, because it is the same if the centre of gravity of the moving body and its displacement are realized in the manner indicated or

---

1. Such was the considered opinion of P. Varignon: 'A motion resulting from the concurrent action of two or more forces is usually called a compounded motion: not that it is so on account of several other motions, but because it results from that concurrence of forces as though from a single one which would be compounded of that which they [the forces] employed by way of action there.'
Whilst relying on the compounding of the velocities of two uniform motions in order to establish the parallelogram rule, using an argument that owes nothing to dynamics, Varignon certainly had also an underlying idea that one could only properly speak of compounding in respect of forces. Descartes had said, in 1618, that he declined to compound motions, for want of being able to see any connection. The difficulty remained essentially still the same at the end of the century.

not. The principle followed here is therefore a property of the centre of gravity of a mass consisting of parts. Whether these parts remain concentrated so as to utilize completely the total applied force, or whether they are dispersed and share the action between each of them, has no effect on the displacement of the centre of gravity.

'That explanation', said Leibniz, 'may serve instead of a proof, for as much progress is made in that way as before, but those who demand a proof *à la façon ordinaire* will readily find it by proceeding as follows. If we draw two straight lines through A so as to make a right angle at A, it will be possible to resolve every one of all those particular motions into two on the sides of that right angle.'

Leibniz, then, considers the property of the centre of gravity, which has just been explained, as fundamental and sufficient in itself. However, he foresaw difficulties on that account for some readers and agreed to put forward a justification. It consists in taking the components of the displacements with respect to rectangular axes. In the text of 1692, which we have just quoted, he takes the case of motions all located in the same plane, and considers two axes only. In the text of 1693, he adds the simple generalization which is indispensable: 'That if the given motions are not in the same plane, then three straight lines at right angles to each other must be used.'

What is the advantage of this analytical method? It is that 'we know that the distance between A and the centre of gravity of the points taken on the same straight line with A is the arithmetical mean of the distances between A and those points, however many they may be.'

$\sum x_i = n . X$, where x_i represents the abscissa of each point of the same 'weight' and n the number of points, is in modern terms the equation which defines the abscissa X of the centre of gravity. If it were not regarded as such at the time when Leibniz was writing, it is nevertheless a geometrical property or a simple consequence of the definition of the centre of gravity, and that is all that matters for our present purpose. The application of co-ordinates to the components of the motions on the axes follows immediately. 'In order to have the distance between A and the point of tendency (of the total motion)' on the axis under consideration, we must 'multiply the distance of the centre of gravity of all the points of tendency on the same side by the number of tendencies'. If x_i represents in fact the projection of each tendency, $\sum x_i$ is the compounding of all the component tendencies aligned on the same axis, and the equation

$\sum x_i = n \cdot X$ gives the means of knowing the first member from the second, that is to say, the required compounding.

The advantage of the analytical method, in the present-day meaning of the word, is, therefore, in the view of Leibniz, that his reader need not assume the dynamic property of the centre of gravity. By making the projection on to the axes, the summations are a matter of course and immediately introduce the co-ordinates of the centre of gravity of the points of tendency as a consequence of the geometric definition of such a centre. Unfortunately, if that proof relieves the reader of a difficulty, it is necessary, in order to accept it, to assume that each particular motion of the assembly which it is required to compound can be itself 'resolved' along the axes of the co-ordinates. Those axes were undoubtedly rectangular and that is possibly the reason why Leibniz had no scruples. We have already pointed out above why, in his time, the compounding or the resolution of motions along rectangular directions seemed readily acceptable. Nevertheless, from a strictly objective point of view, the proof given by Leibniz remains based on a *petitio principii*. It assumes that a particular case has been established from the general result that is sought, and nothing, other than pure and simple guesswork, authorizes the consideration of that particular case as distinct from that which it is really a matter of proving.

Consequently, Leibniz was not very happy in his effort of *captatio benevolentiae* expressly for those who did not move in the same world as himself and who needed proofs '*à la façon ordinaire*'.

Could he have done otherwise? Most probably, no!

It was his meeting with Huygens at Paris that enabled him to familiarize himself with the subject of gravity.[1] He wrote as follows: 'Whilst talking, he [Huygens] realized that I did not have a proper understanding of the centre of gravity. He explained it briefly and added that Dettonville [*i.e.*, Pascal] had given a noteworthy treatment of the subject'. On the information given by Huygens, Leibniz embarked on a reading of Pascal. All Pascal's manuscripts were communicated to him by the brothers Périer. The equation equivalent to that which would nowadays be written as $\sum m_i x_i = m \cdot X$, emerged sufficiently from Pascal's work for Leibniz to have his attention fixed on the form of that result concerning the centre of gravity.

Nevertheless, the idea of dynamical equivalence contained in the

1. Huygens, *Œuvres complètes*, VII, p. 244 and note 12.

concept of centre of gravity (it would be better to say, centre of masses), started to be accepted as fundamental at that time, as is witnessed by the treatise on motive forces by Pardies.[1] It had been launched by Descartes[2] in a way that was still confused and restricted to the field of gravity. It generalized an extremely old intuition in the knowledge of equilibrium. In the same way that for every body there is a point whose stoppage ensures indifferent equilibrium of the body, so also, when there is motion, that point or centre moves as if everything, matter and force, were concentrated there.

We have the feeling that in the case of writers such as Pardies it was a matter there of principle, and consequently that it was not necessary to provide a proof.[3] The situation in which Leibniz found himself when he put forward his general rule for compounding motions explains his irresolution of mind: what is it that is assumed? What is it that has to be proved? We do not know for certain, and the problem varies according to the questioner.

As a result of his study of Pascal, Leibniz was certainly amongst those who were accustomed to include the property of the centre of gravity in the small number of properties essential to an understanding of the motion of bodies.[4] It was quite natural that he should have thought of making use of it when devoting himself to the quest for a general rule for compounding motions. It was quite natural that he should have reverted to those considerations, similar to the ones derived from his study of Pascal, in order to put forward a proof '*à la façon ordinaire*' for those who would have some difficulty in following

1. I. G. Pardies, S.J., *La Statique ou la Science des Forces mouvantes*, Paris, 1673, p. 15.
2. Letter from Descartes to Cavendish, 30 March 1646. *Œuvres de Descartes*, Ed. Adam-Tannery, IV, p. 380 *et seq*. See Pierre Costabel: '*Centre de gravité et équivalence dynamique.*' *Conférences du Palais de la Découverte*, Paris, Sér. D, No. 34, 4 December 1954.
3. See also P. Varignon: *Projet d'une nouvelle Méchanique*, 1687.
'Postulate: In every body that moves or makes an effort to move, there is always a certain point which, encumbered with the impress of all the others, causes the body to accompany that influence which it consequently possesses towards the place whither it tends. There is no need to be troubled on account of that point being the same in all possible positions of the body. It is sufficient, if, in every situation, there be one that we call here its centre of gravity or more generally its centre of direction or of equilibrium, at least for the time that impels the body to follow its impress in that manner.'
4. Letter from Leibniz to Johann Bernoulli, 8/18 March 1698. Gerhardt,*Die phil. Schr. v. Leibniz*, IV, p. 412 and note 2. Leibniz definitely states that when putting his ideas in order at Rome in 1689, after his discussions with Auzout, he wrote a pamphlet in which were proved all those things, concerning force, both absolute and directive, and concerning the *conserved progression of the centre of gravity*. This relates to the text left at Florence with von Bodenhausen.

his argument. In doing so, he merely specified in respect of points of equal mass a type of proof with which he was familiar in respect of points of any mass. We have already said that he might not have realized that he was begging the question by doing so; consequently, it would seem that we should not be too hard upon him.

There remains another more serious objection that could be levelled at the present time against the argument used by Leibniz. When he states that one quarter of the moving body must go four times further in obedience to the same 'tendency', is not he acting like a Cartesian and accepting the conservation of quantity of motion? Nevertheless, we have already definitely pointed out above that it is not a question here of a scalar quantity of motion, which would be in effect Cartesian, but of a quantity of motion considered with the sense and direction of motion—a quite different matter.

In the version of 1693, it is most noteworthy that Leibniz adds: 'It is well to note that in the compounding of motions, the same quantity of progression is conserved, and not always the same quantity of motion. For example, if there are two tendencies in the same straight line, but in contrary senses, the moving body goes in the *stronger* direction with the difference of the velocities and not with their sum, as would happen if the tendencies took effect from the same side. If the two contrary tendencies were equal, there would be no motion.'

So, in his last version, Leibniz felt the necessity of defining more closely yet again that he did not regard matters as did the Cartesians. The term quantity of motion he leaves to them; he considers the quantity of progression, which is nothing more than the quantity of motion taken in the vectorial sense. Furthermore, he gives an account of the argument underlying his thought. Projections on the axes, we should now say the use of analytical geometry, allow that consideration to be justified. Projection on to an axis gives in effect the algebraic sum of the tendencies by which the overall action is characterized. If it be zero, then there is no resultant motion. If it be not zero, then it defines the sense and the magnitude of the resultant motion. With regard to establishing those facts, it is the algebraic sums of the quantities of progression that are characteristic.

However, it will be said, that when Leibniz, applying his mind to the moving body divided into parts, states that one quarter of the moving body must go four times further in obedience to the same tendency, everything takes place as if he regarded the quantity of progression as equivalent to or representative of the force intimated by

101

the tendency. That differs from Descartes only by the addition of the sense and the direction of the scalar velocity. In short, is not the Leibnizian force $\overrightarrow{mv}$?

In the text under consideration, Leibniz continues by saying: 'Nevertheless that suffices so to speak *in abstracto*, when we *already* assume those tendencies to be present in the moving body; but *in concreto*, in considering the causes that must produce them there, we shall find that not only is the same quantity of progression conserved in all, but also the same quantity of absolute and total force which is yet different from the quantity of motion.'

Thus, we should be wrong to believe that Leibniz confuses force and quantity of progression. The former is a fundamental reality which comes into play always when we consider things *in concreto*. The latter suffices to take account of the phenomenon *in abstracto*.

Seeing that M. Guéroult has given a perfect explanation of this difficult and rather obscure difference, as it appears to us at the present day, we cannot do better than follow his account in its broad outlines.

In the texts quoted in the previous chapter, Leibniz seems to identify *conatus* with potential force on the one hand, and impetus with kinetic force on the other hand, the former being compared with the latter as 'the point to the line', that is to say, as a differential element to an integral. Nevertheless, impetus is defined as quantity of motion,[1] and it is definitely stated in the second *Essay de Dynamique* that the expression *mv* 'is valid only in the case of potential force or of infinitesimal motion which I am accustomed to call solicitation'.[2] Consequently, impetus could not be confused with kinetic force. That, moreover, is what Leibniz explicitly states: 'However, although impetus is always connected with kinetic force, there is a distinction between them.'[3] How could it be otherwise, seeing that the expression for kinetic force is mv^2? All the same, how do we escape from the apparent confusion in the Leibnizian ideas?

'The fact is', says Guéroult, 'that impetus may be considered from two points of view.[4]

'(*a*) By itself, at the moment when it is produced, *neglecting* the kinetic force of which it is the effect. In that moment there is movement of a certain mass along a certain path; and the expression for it

1. *Dynamica de potentia.* Gerhardt, *Leibn. math. Schr.*, VI, p. 398.
2. Gerhardt, *Leibn. math. Schr.*, VI, p. 218.
3. *Specimen Dynamicum.* Gerhardt, *ibid.*, VI, p. 238.
4. M. Guéroult, *Dynamique et Métaphysique Leibniziennes*, Paris, 1934, p. 41.

is definitely mv seeing that it is a question of *uniform* motion without acceleration considered by itself in that moment, and the distance traversed is exactly proportional to the velocity. The case is comparable with that of potential force, seeing that the latter relates to an infinitesimal moment of motion, considered outside of any temporal succession, independently of any process of accumulation. Therefore, it is natural for the same expression mv to be valid for one and the other.

'(*b*) However there is a difference, and it is that the moment considered in the case of potential force being as it were the originating one, the quantity of motion is consequently simply embryonic, whereas with impetus, the moment, although considered by itself, nevertheless carries within it the result of an earlier accumulation of motions. . . .

'Impetus has, therefore, two different aspects: an external aspect by which it is recognized *in abstracto*, in an isolated moment, and opposes the kinetic force so as to approximate to the potential force through its expression mv; an internal aspect by which it is related to its cause, and the moment which it represents is no longer considered in isolation, but is linked with its genesis and as the result of a multitude of earlier moments which it embodies and which confer its distinguishing mark. It seems, then, to be closely linked with kinetic force and opposed to potential force, as being able to represent the former adequately and to replace it, for it is born and dies at the same time as the kinetic force. Furthermore, that which completely destroys the impetus also consumes the kinetic force and evaluates it, namely, work whose expression is mv^2.'

We thought it desirable to quote the above in its entirety, because it explains perfectly the crucial question to which we had come.

We can quite understand why Leibniz carefully avoided speaking of force in the text with which we are concerned, and systematically used the term 'tendency'. These 'tendencies', which we can assume to be '*already*' in the moving body *in abstracto*, are not forces. They exist because there are motions (and that explains how we may treat them as equivalent in the discourse). Their existence is linked *in concreto* with absolute forces, that is to say, with kinetic forces, and manifests rather than represents those forces. However, if we detach ourselves from the concrete aspect in order to consider *in abstracto* what happens in an isolated moment, then those tendencies represent a force capable of acting, whose measure is that of impetus given by

103

the expression *mv*. It was not by chance that in the additions to the text of 1693 for publication in the *Journal des Sçavans*, Leibniz indicated that the motions resulting from the tendencies whose compounding was under consideration are *uniform* motions. In the isolated moment, viewed abstractly, that uniformity approximates to the tendency of the potential force, the motion of an infinitesimal movement[1] and lays the foundation of its evaluation by the product of the mass and the velocity.

The analysis made by M. Guéroult of the Leibnizian texts is therefore confirmed by the *Règle générale de la composition des mouvements* of 1692–1693. When making further explanation at a later date, Leibniz made no change in the essential features of his system.

The final argument in the method adopted to establish the *Règle générale* seems just as clear. Leibniz could not treat the problem of compounding motions otherwise than as a problem in dynamics, and treat it as a function of principles other than those of that science. Through the supreme intelligibility that he had found in the realism of force, in conjunction with other realms of thought, it was necessary that 'his' science—(that was what he called it)—it was necessary that 'dynamics' take account in particular of that compounding of motions, the logical difficulty of which, users such as Lamy and Varignon, had no suspicion.

Conclusion

In copying the two texts which we have been considering, was Des Billettes aware of their close and intimate connection? The unknown person who gave these two texts the same fate in the midst of anonymous papers in the archives of the *Académie*, did he suspect anything? Undoubtedly, we shall never know, but the fortunate conjunction of circumstances that has provided us with such a valuable lesson none the less merits our gratitude to its benevolent contrivers.

If we have been obliged to go beyond the limits of dynamics in order to study solutions to a geometrical problem on the construction of tangents, we do not regret it. Apart from the intrinsic interest of those solutions, we have received a lesson in hidden connections and unsuspected pathways in thought between domains of knowledge

1. See chapter II, pp. 61–62.

which our present-day disciplines have accustomed us to distinguish rather carefully. What surprise was in store for us! In short, the Leibnizian solution bringing us back to the main object of this discussion, namely, an essay on the rational constitution of a knowledge of motion.

We have not spared criticism in the course of this essay. In particular, the illusion of being able to evade the essential difficulty of the concept of time by eliminating it to the advantage of reality in dynamics or energetics. Who could say that the very inadequacies, thus brought to light, are not instructive? Simplicity is always a semblance; understanding of the true nature of phenomena is always a hazardous undertaking.

That which was lacking in the analysis provided by Descartes and by Leibniz, namely, consideration of time *qua* dimension *sui generis*, classical science believes it has discovered in the concept of a temporal, universal, independent and indifferent framework within which events of this universe unfold themselves as though in a stage setting. Certainly, the separation of concepts and their distinct treatment have been beneficial, and remain the principle of a healthy scientific method. However, we know sufficiently well since Relativity that the time of classical mechanics is but a provisional scheme, whose worth has been taken up anew even in extremely remote fields, such as palaeontology, in connection with the evolution of species. After having separated and distinguished, it is then necessary to come back to unifying visions of things.

Simplicity is always a semblance, we said above. It is obviously useful that we have been able to receive a lesson to that effect from the enlightening intelligence of Leibniz, who was eager to grasp everything for fear of allowing the most tenuous thread belonging to the structure of this universe to escape; he was prey to a prodigious activity and not to sterile erudition or distinguished dilettantism. Again, it is useful that we have been able to put our finger on the penalty of material errors resulting from that prodigious activity. Genius is more human and more sympathetic when it contributes its share to the common frailty of mankind, when it, too, is subjected to the limitations imposed by a strained intellectual effort.

However, much more can be said here and the above reflections suggest what it might be. The story that we have tried to tell inspires admiration and respect. Admiration for the coherence of a mind, however difficult, complex and subtle it may be, whatever its weaknesses. Respect for the purpose pursued without weakening in outlook.

The elements of a positive heritage for science have emerged from the unswerving will of internal coherence, and in that respect Leibniz, like many others, remains a very great master.

Appendix I

ESSAY DE DYNAMIQUE

1. Définition

De la force égale, moindre, et plus grande.

Lorsqu'il y a deux états tellement faits que si l'un pouvait être substitué à la place de l'autre sans aucune action du dehors, il s'ensuivrait un mouvement perpétuel mécanique, on dira que la force aura été augmentée par cette substitution, ou que la force de l'état substitué sera plus grande, et que celle de l'état pour lequel on l'a substitué était moindre; mais que si la force est ni moindre ni plus grande elle est égale.

Scholie

J'appelle ici état, (statum) *un corps ou plusieurs pris avec certaines circonstances de situation, de mouvement, etc. J'ai voulu me servir de cette marque extérieure de la force augmentée qui est la réduction au mouvement perpétuel pour m'accommoder davantage aux notions populaires, et pour éviter les considérations métaphysiques de l'effet et de la cause. Car pour expliquer les choses,* a priori, *il faudrait estimer la force par la quantité de l'effet prise d'une certaine manière qui a besoin d'un peu plus d'attention pour être bien entendue. Mais comme ce discours préparera le lecteur, je ne laisserai pas de faire entrer en passant quelques considérations de la cause et de l'effet.*

2. Définition

La quantité du mouvement est le produit de la masse du corps par sa vitesse.

Appendix I

The text in French is taken from the copy made by Des Billettes. Modern spelling has been used; the figure has been put in its proper place; punctuation and the writing of numbers have been retained, in which respect the copy agrees with the original manuscript at Hanover, except in one or two places. The general accuracy of the copy is revealed by the footnotes which give the variations present in the original.

ESSAY ON DYNAMICS

1. Definition

On equal, less, and greater force.

When two states are so constituted that if it were possible to replace one by the other without the intervention of any external action, then perpetual mechanical motion would result; and it can be said that the force will have been increased by that substitution, or that the force of the state which is substituted will be greater, and that the force of the state for which it has been substituted was less; but if the force is neither less nor greater, then it is equal.

Scholium

I give here the name of *state* (*statum*) to one or more bodies under certain conditions of position, motion, etc. I wished to make use of that outward sign of increased force which reduces to perpetual motion in order to adapt myself more to popular ideas, and in order to avoid metaphysical considerations of effect and cause. For in order to explain matters, *a priori*, it would be necessary to evaluate the force by the quantity of the effect considered in a certain way which has need of a little more attention in order to be properly understood. However, as this discourse will prepare the reader, I shall not refrain from introducing, in passing, some considerations on cause and effect.

2. Definition

The quantity of motion is the product of the mass of a body and its velocity.

Scholie

La masse des corps sensibles s'explique par la pesanteur. Ainsi un corps étant de 4 livres et allant avec un degré de vitesse, il aura une quantité de mouvement comme quatre. *Mais si étant de 4 livres il avait 3 degrés de vitesse, sa quantité de mouvement serait comme 12.*

3. Définition

Le mouvement perpétuel mécanique (qu'on demande en vain) est un mouvement où les corps se trouvent dans un état violent, et agissant pour en sortir n'avancent pourtant point, et le tout se retrouve au bout de quelque temps dans un état non seulement autant violent que celui où l'on était au commencement, mais encore au delà, puisque outre que le premier état est restitué il faut que la machine puisse encore produire quelque effet ou usage mécanique, sans qu'en tout cela aucune cause de dehors y contribue.

Scholie

Par exemple, il y a une machine dans laquelle au commencement quelques poids étaient élevés à une certaine hauteur. Ces poids se retrouvant dans un état violent font effort pour descendre, et il y en a qui descendent effectivement, et qui obligent d'autres à monter. Mais la nature se trompe (pour parler ainsi) en croyant d'arriver à son but, et l'art ménage si bien les choses qu'au bout de quelque temps il se trouve qu'il y a tout autant de poids élevés qu'au commencement, et même au delà. Je dis, au delà, *puisque chemin faisant ces poids ont encore pu avoir et faire quelqu'autre effet violent, par exemple élever de l'eau, moudre du blé, ou produire quelqu'autre chose selon l'usage auquel on a destiné la machine. Un tel mouvement perpétuel a toujours été cherché; mais il est impossible de le trouver car la force augmenterait d'elle-même et l'effet serait plus grand que la cause totale. Il est vrai que si l'on ôte les empêchements accidentels les corps descendants peuvent remonter précisément d'eux-mêmes à la première hauteur. Et cela est nécessaire; autrement la même force ne se conserverait pas, et si la force se diminue l'effet entier n'est pas équivalent à la cause, mais inférieur. On peut donc dire qu'il y a un mouvement perpétuel physique, tel que serait un pendule parfaitement libre; mais ce pendule ne passera jamais la première hauteur, et même il n'y arrivera pas s'il opère ou produit le moindre effet en son chemin, et s'il surmonte le moindre obstacle; autrement ce serait un mouvement perpétuel mécanique. Or ce qu'on on vient de dire des poids a lieu aussi à l'égard des ressorts et autres corps qu'on fait agir en les mettant dans un état violent.*

Scholium

The mass of tangible bodies is explained by gravity. Thus, a body being of 4 pounds and moving with one degree of velocity, it will have a quantity of motion as *four*. But if being of 4 pounds it had 3 degrees of velocity, its quantity of motion would be as 12.

3. Definition

Mechanical perpetual motion (which is sought in vain) is a motion in which bodies find themselves in a violent state, and straining to get out of it nevertheless make no advance, and after some time everything is once more in a state not only just as violent as it was in the beginning, but even more so, because apart from the fact that the first state is restored the machine must still be able to produce some effect or do some mechanical service without any external cause contributing anything thereto.

Scholium

For example, there is a machine in which at the start some weights were raised to a certain height. Those weights finding themselves in a violent state make an effort to fall down, and some of them do effectively fall down, and compel others to rise. However, nature is deceived (so to speak) in believing it possible to achieve its purpose, and art arranges matters so well that at the end of a certain time it happens that there are just as many weights in the raised position as at the beginning, and even more. I say, *even more*, because those weights on their way have still been able to receive and to perform some other violent effect, for example, to raise water, grind corn, or perform something else depending on the use for which the machine was intended. Such perpetual motion has always been sought after; but it is impossible to find it for the force would increase of itself and the effect would be greater than the total cause. It is true, that if we remove the accidental hindrances the falling bodies can ascend by themselves to the original height. And that is necessary; otherwise the same force would not be conserved, and if the force decreases, the whole effect is not equivalent to the cause, but is less. We can say therefore that there is a physical perpetual motion, as would be the case of a perfectly free pendulum; but that pendulum will never go beyond the original height, and it will not even reach that height if it brings about or produces the least effect in its path, or if it overcomes the least obstacle; otherwise that would be mechanical perpetual motion. Now, what has just been said about weights occurs also in respect of springs and other bodies which are made to act by putting them into a violent state.

111

Axiome 1

La même quantité de la force se conserve, ou bien, l'effet entier est égal à la[1] cause totale.

Scholie

Cet axiome est d'aussi grand usage pour la mécanique, que celui qui dit que le tout est égal à toutes ses parties prises[2] ensemble, *est utile dans la géométrie; l'un et l'autre nous donnant moyen de venir à des équations; et à une manière d'analyse. Il s'ensuit qu'il n'y a point de mouvement perpétuel mécanique, et même qu'il n'arrivera jamais que la nature substitue un état à la place de l'autre s'ils ne sont d'une force égale. Et si l'état L se peut substituer à la place de l'état M il faut que réciproquement l'état M se puisse substituer à la place de l'état L sans crainte du mouvement perpétuel, par la définition de la force égale ou inégale, que nous avons donnée.*

Axiome 2

Il faut autant de force pour élever une livre à la hauteur de 4 pieds qu'il en faut pour élever 4 livres à la hauteur d'un pied.

Scholie

Cet axiome est accordé. On le pourrait démontrer néanmoins par l'axiome 1° et autrement. Et sans cela il serait aisé d'obtenir le mouvement perpétuel.

Postulatum ou demande 1

On demande que toute la force d'un corps donné puisse être transférée sur un autre corps donné, ou du moins, si on suppose cette translation, qu'il n'en arriverait aucune absurdité.

Scholie

Il est sûr qu'un petit corps peut acquérir une telle vitesse qu'il surpasse[3] la force d'un grand corps qui va lentement. Il pourra donc l'acquérir[4] précisément égale. Et le grand corps en pourra être la cause en perdant sa force par des actions sur d'autres corps qui enfin la pourront transférer toute sur le seul petit par des rencontres ou changements propres à cela. De même le petit corps pourra transférer toute la sienne sur le grand corps, et il

État de l'autographe:
1. ... *l'effet entier est égal à* sa *cause totale.*
2. ... *le tout est égal à toutes ses parties ensemble.*
3. ... *qu'il surpassera ....*
4. ... *il pourra donc l'acquérir* aussi *précisément égale.*

Axiom 1

The same quantity of force is conserved, or rather, the whole effect is equal to the[1] total cause.

Scholium

This axiom is of just as much use for mechanics as that which says that *the whole is equal to all its parts taken together*[2] is useful in geometry; the one and the other giving us the means of arriving at equations; and at a method of analysis. It follows that there is no mechanical perpetual motion, and in fact that it will never happen that nature substitutes one state for another if they are not of equal force. And if the state *L* can be substituted for the state *M*, then reciprocally the state *M* must be capable of being substituted in place of the state *L* without fear of perpetual motion, by definition of equal or unequal force, which we have given.

Axiom 2

As much force is required to raise one pound to a height of 4 feet as to raise 4 pounds to a height of one foot.

Scholium

This axiom is conceded. Nevertheless, it would be possible to prove it by axiom 1, or otherwise. And without that it would be easy to obtain perpetual motion.

Postulate 1

It is postulated that all the force of a given body can be transferred to another given body, or at least, if we assume that transfer, that no absurdity would result therefrom.

Scholium

It is certain that a small body can acquire such a velocity that it surpasses[3] the force of a large body which moves slowly. Therefore, it will be able to acquire[4] an exactly equal force. And the large body will be able to be the cause of it by losing its force through actions on other bodies which will finally be able to transfer it all to the single small one by encounters or changes suitable thereto. In the same way the small body will be able to transfer all its force to the large body, and it is of

Original manuscript has:
1. ... the whole effect is equal to *its* total cause.
2. ... the whole is equal to all its parts together.
3. ... that it *will surpass* ....
4. ... Therefore, it will be able to acquire an exactly equal force *also*.

n'importe pas si cela arrive médiatement ou immédiatement, tout d'un coup ou successivement, pourvu qu'au lieu que d'abord le seul corps A était en mouvement, il se trouve à la fin que seul le corps B est en mouvement. Car ainsi il faut bien qu'il ait reçu toute la force du corps A par l'axiome 1° autrement une partie en serait périe. On peut imaginer certaine machination pour l'exécution de ces translations de la force. Mais quand on n'en donnerait pas la construction, c'est assez qu'il n'y ait point d'impossibilité, tout comme Archimède prenait une droite égale à la circonférence d'un cercle sans la pouvoir construire.

Demande 2

On demande que les empêchements extérieurs soient exclus ou négligés, comme s'il n'y en avait aucuns.

Scholie

Car puisqu'il s'agit ici du raisonnement pour estimer les raisons des choses et nullement de la pratique, on peut concevoir le mouvement comme dans le vide, afin qu'il n'y ait point de résistance du milieu, et on peut s'imaginer que les surfaces des plans et des globes sont parfaitement unies, afin qu'il n'y ait point de frottement, et ainsi du reste.[1] C'est afin d'examiner chaque chose à part sauf à les combiner dans la pratique.

Proposition 1

Lemme démontré par d'autres

Les vitesses que les corps pesants acquièrent en descendant sont comme les carrés[2] des hauteurs dont ils descendent, et vice versa, les corps en vertu des vitesses qu'ils ont peuvent monter aux hauteurs dont ils devraient descendre pour acquérir ces vitesses.

Scholie

Cette proposition a été démontrée par Galiléi, Mr Hugens et autres. Par exemple si un corps descendant de la hauteur d'un pied acquiert au bout de la chute un degré de vitesse, un corps descendant de 2 pieds acquerra 4 degrés de vitesse. 3 pieds donneront 9 degrés, 4 pieds 16 degrés, etc. Car 4, 9, 16 sont les nombres carrés de 2, 3, 4, etc. et vice versa, un corps d'un degré de vitesse pouvant monter à la hauteur d'un pied, il s'ensuit

---

1. . . . *et ainsi* des autres.
2. *Le texte copié par Des Billettes concorde bien ici avec l'original de Hanovre. Leibniz a écrit effectivement:* « comme les carrés des hauteurs », *alors qu'il aurait dû dire:* « comme les racines carrées ». *Le début du Scholie montre que Leibniz poursuit son développement en suivant cette donnée erronée.*

no consequence if that happens directly or indirectly, suddenly or successively, provided that initially only the body *A* was in motion, and that finally only the body *B* is in motion. For it must of necessity have received all the force of body *A* in that manner according to axiom 1, otherwise some part of it would be destroyed. We can imagine a certain contrivance for carrying out these transfers of force. But even though we should not describe its construction, it is enough that there should be no impossibility in doing so, just as Archimedes took a right angle at the circumference of a circle without being able to construct it.

Postulate 2

It is postulated that external hindrances are excluded or neglected, as though there were none.

Scholium

For, seeing that it is a question here of the line of argument for the purpose of evaluating the grounds of the case and not the application, we may regard the motion as taking place in a void, so that there would be no resistance from the medium, and we may imagine that the surfaces of the planes and spheres are perfectly smooth, in order that there may be no friction, and so on.[1] That is for the purpose of examining each thing by itself, but to combine them in practice.

Proposition 1

Lemma proved by others

The velocities that bodies acquire in falling are as the squares[2] of the heights from which they fall, and *vice versa*, the bodies in virtue of the velocities which they have are able to rise to the heights from which they must have fallen in order to acquire those velocities.

Scholium

This proposition has been proved by Galileo, Mr Huygens and others. For example, if a body falling from the height of one foot acquires at the end of the fall one degree of velocity, a body falling from 2 feet will acquire 4 degrees of velocity. 3 feet will give 9 degrees, 4 feet 16 degrees, etc. For 4, 9, 16 are the squares of 2, 3, 4, etc., and *vice versa*, a body

Original manuscript has:
1. ... and so *in other respects.*
2. The text copied by Des Billettes is in good agreement at this point with the original at Hanover. Leibniz has in fact written—'*comme les carrés des hauteurs*' (as the squares of the heights)—whereas he ought to have said—'*comme les racines carrées*' (as the square roots). The opening of the Scholium shows that Leibniz continued his development by following this erroneous statement.

qu'un corps de 4 degrés de vitesse aura la force de s'élever à 16 pieds.[1] *En tout ceci il n'importe point si le corps est grand ou petit, ni si sa descente se fait perpendiculairement ou obliquement, pourvu qu'on observe la deuxième demande. Cependant en estimant la hauteur on entend toujours la hauteur perpendiculaire.*

Proposition 2

Un corps A pesant une livre et descendant de la hauteur de 16 pieds peut élever un corps B pesant 4 livres à la hauteur qui soit tant soit peu moindre que de 4 pieds.

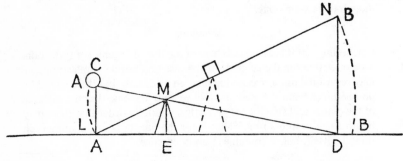

Figure 15

Démonstration

Cela se prouve aisément par la statique commune. Concevons une balance à bras inégaux LMN dont le centre,[2] *et le bras MN soit un peu plus que 4 fois de la longueur du bras LM. Cette balance soit obliquement située, en sorte que le bout L soit dans l'horizon, et arrive au poids A, et que le bout M arrive au poids B élevé à 16 pieds. Cela étant il est manifeste que si ces bouts de la balance sont engagés à ces poids et les soutiennent, B prévaudra à A par le principe vulgaire de l'équilibre. Car comme A est quadruple de B, si MN était aussi quadruple de ML, tout serait en équilibre.*

1. *Ici Leibniz corrige et rétablit la vérité. L'examen de l'original de Hanovre montre qu'il a d'abord écrit que le corps de 4 degrés de vitesse peut remonter à 2 pieds, puis il a rayé pour mettre 16 pieds. Les choses s'expliquent donc aisément si l'on admet* une rédaction rapide. *Par défaut d'attention Leibniz écrit: « carrés des hauteurs » au lieu de « racines carrées », il poursuit vite, ce n'est qu'arrivé à la réciproque qui lui importe le plus pour ses raisonnements qu'il s'aperçoit de l'erreur, il corrige, mais ne prend pas le temps de revenir en arrière pour corriger l'ensemble.*
État de l'autographe
2. *... dont le centre* soit M, *et le bras...*

with one degree of velocity being able to rise to the height of one foot, it follows that a body with 4 degrees of velocity will have the force to raise itself to 16 feet.[1] In all this, it matters not if the body be large or small, nor if its fall be made perpendicularly or obliquely, provided that the second postulate be observed. Nevertheless, in evaluating the height the perpendicular is to be understood always.

Proposition 2

A body *A* weighing one pound and falling from a height of 16 feet is able to raise a body *B* weighing 4 pounds to the height which is somewhat less than 4 feet.

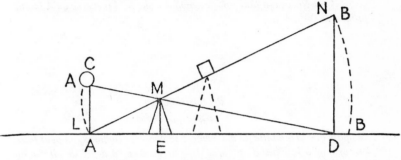

Figure 15*

Proof

That is easily proved by ordinary statics. Imagine a balance with unequal arms *LMN* whose point of support[2] [is *M*] and the arm *MN* is a little more than 4 times the length of the arm *LM*. Let the balance be placed obliquely so that the end *L* is at ground level and receives a

---

* The original manuscript has two similar figures in the margin. The first, reduced, at the side of a paragraph which has been crossed out. The second, lower down, has been corrected by Leibniz who had drawn the arm *MN* too long. Consequently, there are crossings out and redrawing of *NDBB* nearer to *M*. The copyist did not understand and Des Billettes has reproduced a meaningless item between *M* and *N*. The figure appears at the end of the version by Des Billettes.

1. At this point Leibniz corrected himself and restored accuracy. Examination of the original at Hanover shows that he wrote in the first place that the body of 4 degrees of velocity is able to rise to 2 feet, then he crossed it out in order to put 16 feet. The matter is easily explained if we assume *rapid composition*. Through lack of attention Leibniz wrote '*carrés des hauteurs*' instead of '*racines carrées*'; it was only when he came to the reciprocal statement, which is more important for his argument, that he noticed the mistake; he corrected it, but did not take the time to go back in order to correct the whole.

Original manuscript has:

2. ... whose point of support *is M*, and the arm ...

Mais si MN étant tant soit peu plus que quadruple de ML, B l'emportera et descendant jusqu'à B dans l'horizon, il fera monter A jusqu'à A. Maintenant des points AMB menons des perpendiculaires sur l'horizon, savoir AC, ME, BD. Or ME est à BD comme LM est à LN et par hypothèse LN est tant soit peu plus que le quintuple de ME. Or ME est à AC comme MB à AB c'est-à-dire comme MN à LN et MN est à LN en raison tant soit peu plus grande que 4 à 5. Et par conséquent encore ME est ainsi à AC. Donc BD étant à ME en raison tant soit peu plus grande que de 5 à 1 et ME étant à AC en raison tant soit peu plus grande que de 4 à 5, il s'ensuit que BD sera à AC en raison tant soit peu plus grande que de 4 à 1. C'est-à-dire BD sera un peu plus que le quadruple de AC et par conséquent BD étant de 16 pieds, par hypothèse, il est manifeste que AC hauteur à laquelle le corps A est élevé sera tant soit peu moindre que de 4 pieds, ce qu'il fallait prouver.

Proposition 3

Problème

Supposé que la quantité de mouvement se conserve toujours, on peut faire en sorte qu'à la place d'un corps de 4 livres avec un degré de vitesse on obtienne un corps d'une livre avec 4 degrés de vitesse.

Démonstration

Car le 1er corps soit A, le second B et soit toute la force d'A transférée sur B (demande 1) c'est-à-dire qu'au lieu qu'A était seul en mouvement, soit maintenant B seul en mouvement, rien d'accidentel ou d'extérieur ayant absorbé quelque chose de la force (demande 2), il faut que B ait la même quantité de mouvement qu'A (par l'hypothèse). Donc A de 4 livres ayant eu une vitesse d'un degré (par l'hypothèse) il faut que B qui est d'une livre (par l'hypothèse) reçoive la vitesse de 4 degrés. Car ce n'est qu'ainsi que B aura la même quantité de mouvement qu'A suivant la définition 2 puisque une livre de B doit être multipliée par 4 degrés pour faire autant que les 4 livres d'A multipliées par un degré, ce qu'il fallait faire.

Proposition 4

Problème

Supposé qu'à la place de 4 livres avec un degré de vitesse on puisse

118

weight *A*, and so that the end *N* receives a weight *B* raised to a height of 16 feet. That being so, it is manifest that if those ends of the balance be connected with those weights and support them, then *B* will prevail over *A* by the common principle of equilibrium. For, as *A* is four times *B*, if *NM* were also four times *ML*, everything would be in equilibrium. But, if *MN* being somewhat more than four times *ML*, *B* will prevail and, falling to *B* at ground level, it will cause *A* to rise up to *A*. Now, drop perpendiculars *AC*, *ME*, *BD* from the points *AMB* to the ground. Now, *ME* is to *BD* as *LM* is to *LN*, and by hypothesis *LN* somewhat more than five times *ME*. Now, *ME* is to *AC* as *MB* is to *AB*, that is to say, as *MN* to *LN*, and *MN* is to *LN* in the ratio somewhat greater than 4 to 5. And consequently so is *ME* to *AC*. Therefore, *BD* being to *ME* in a ratio somewhat greater than 5 to 1 and *ME* being to *AC* in a ratio somewhat greater than 4 to 5, it follows that *BD* will be to *AC* in a ratio somewhat greater than 4 to 1. That is to say, *BD* will be a little more than four times *AC* and therefore *BD* being at a height of 16 feet, by hypothesis, it is manifest that *AC*, the height to which the body *A* is raised, will be somewhat less than 4 feet. *Q.E.D.*

Proposition 3

Problem

It being assumed that the quantity of motion is always conserved, we can arrange that instead of a body of 4 pounds with one degree of velocity we have a body of one pound with 4 degrees of velocity.

Proof

Let the first body be *A*, the second body be *B* and let all the force of *A* be transferred to *B* (Postulate 1), that is to say, instead of *A* alone being in motion, *B* alone is now in motion, nothing accidental or external having absorbed anything of the force (Postulate 2), *B* must then have the same quantity of motion as *A* (by hypothesis). Therefore, *A* of 4 pounds having had a velocity of one degree (by hypothesis) *B* which is of one pound (by hypothesis) must acquire a velocity of 4 degrees. For only in that manner will *B* have the same quantity of motion as *A* in accordance with definition 2, seeing that one pound of *B* must be multiplied by 4 degrees in order to make as much as 4 pounds of *A* multiplied by one degree. *Q.E.F.*

Proposition 4

Problem

It being assumed that instead of 4 pounds with one degree of velocity

acquérir une livre avec 4 degrés de vitesse, je dis qu'on pourra obtenir le mouvement perpétuel mécanique.

Démonstration

Faisons qu'un globe A d'une livre[1] de poids descende de la hauteur d'un pied et acquière un degré de vitesse. Soit maintenant obtenu qu'à la place un globe B d'une livre ait 4 degrés de vitesse par l'hypothèse, ce globe B pourra monter à la hauteur de 16 pieds (proposition 1) et puis engagé à une balance qu'il rencontrerait au bout de la montée, et descendant derechef de cette hauteur jusqu'à l'horizon, il pourra élever A à une hauteur tant soit peu moindre que 4 pieds (proposition 2). Or au commencement le poids A se trouvait élevé sur l'horizon d'un pied, et B en repos dans l'horizon. Maintenant il se trouve que B redescendu est encore en repos dans l'horizon, mais qu'A est élevé sur l'horizon presque de 4 pieds (bien au delà de son 1ᵉʳ état, et nous avons le 2ᵉ état, où l'effet est plus grand que la cause, ce qui peut faire le mouvement perpétuel mécanique). Ainsi A avant que de retourner de la hauteur de 4 pieds à sa 1ʳᵉ hauteur d'un pied, pourra faire quelque effet mécanique chemin faisant (élever de l'eau, moudre du blé, etc.), et néanmoins étant retourné à A toutes choses seront restituées au 1ᵉʳ état (et ce jeu pourra continuer toujours) et c'est obtenir le mouvement perpétuel mécanique (définition 3) ce qu'il fallait faire.

Proposition 5

Problème

Supposé que la quantité de mouvement se conserve toujours, on peut obtenir le mouvement perpétuel mécanique.

Démonstration

Car la quantité de mouvement étant toujours conservée (par l'hypothèse) on peut à la place de livres 4, vitesse 1, substituer livre 1, vitesse 4 (proposition 3) et cela étant on pourra obtenir le mouvement perpétuel mécanique (proposition 4) ce qu'il fallait faire.

Proposition 6

Un corps de 4 livres de poids et d'un degré de vitesse a seulement le quart de la force d'un corps d'une livre de poids et de 4 degrés de vitesse.

---

1. *Errenr analogue aux précédentes. Par inattention Leibniz écrit une livre au lieu de quatre, alors que le texte même de la proposition 4 ne laisse aucun doute à ce sujet.*

we can arrive at one pound with 4 degrees of velocity, I say that we shall be able to obtain mechanical perpetual motion.

Proof

Let a sphere A of weight one pound[1] fall from a height of one foot and acquire one degree of velocity. Now, instead, let a sphere B of one pound have 4 degrees of velocity by hypothesis; this sphere B will be able to rise to a height of 16 feet (Proposition 1) and then coming to an equilibrium which would be reached at the end of the rise, and falling anew from that height to the ground, it will be able to raise A to a height somewhat less than 4 feet (Proposition 2). Now, at the start, the weight A was raised one foot above the ground, and B was at rest on the ground. It now happens that B having fallen again is still at rest on the ground, but that A is raised nearly 4 feet above the ground (far beyond its 1st state, and we have the 2nd state, in which the effect is greater than the cause, which can produce mechanical perpetual motion). Consequently, A, before returning from the height of 4 feet to its 1st height of one foot, will be able to perform some mechanical effect as it goes along (raise water, grind corn, etc.), and nevertheless having returned to A everything will be restored to the 1st state (and that activity will be able to continue for ever) and that means achieving mechanical perpetual motion (Definition 3). *Q.E.F.*

Proposition 5

Problem

It being assumed that the quantity of motion is always conserved, then we can obtain mechanical perpetual motion.

Proof

The quantity of motion being always conserved (by hypothesis), instead of pounds 4, velocity 1, we can substitute pounds 1, velocity 4 (Proposition 3) and that being so we shall be able to obtain mechanical perpetual motion (Proposition 4). *Q.E.F.*

Proposition 6

A body of 4 pounds weight and one degree of velocity has only one quarter of the force of a body of one pound weight and 4 degrees of velocity.

1. A similar mistake to the earlier ones. Leibniz has inadvertently written one pound instead of four, whereas the text of proposition 4 leaves no doubt on the matter.

Démonstration

Car soit le premier poids A, et le second B, et supposons qu'A puisse monter à une certaine hauteur, par exemple d'un pied. B pourra monter à la hauteur de 16 pieds (proposition 1) donc B a la force d'élever une livre, savoir son propre poids, à la hauteur de 16 pieds. Et par conséquent (axiome 2) il a la force d'élever 16 livres à un pied au lieu qu'A a seulement la force d'élever 4 livres, c'est-à-dire son propre poids, à 1 pied (par l'hypothèse). Or la force d'élever 16 livres à un pied est quadruple de la force d'élever 4 livres à 1 pied (par le sens commun), donc la force de B est quadruple de la force d'A, ce qu'il fallait prouver.

Proposition 7

Un corps de 4 livres de poids et de un degré de vitesse a la même force qu'un corps d'une livre de poids et de deux degrés de vitesse. Et par conséquent si toute la force de celui-là doit être transférée sur un corps d'une livre, il ne recevra que deux degrés de vitesse.

Démonstration

Soit le premier A, le second B. Si A peut monter à 1 pied, ou élever 4 livres, c'est-à-dire son poids, à 1 pied, B pourra monter à 4 pieds (proposition 1) ou élever son poids qui est de 1 livre à 4 pieds. Donc (axiome 2) la force d'A est égale à celle de B, ce qu'il fallait démontrer.[1]

Scholie

Ces deux propositions se peuvent encore démontrer indépendamment de l'axiome 2, par le seul axiome 1 joint à la définition 1, en employant un mécanisme semblable à celui de la proposition 2 pour réduire celui qui dirait le contraire au mouvement perpétuel mécanique. Aussi avons-nous remarqué à l'axiome 2 qu'on le peut démontrer par l'axiome 1, c'est-à-dire réduisant le contraire au mouvement perpétuel, ou ad absurdum. *Il est bon aussi de remarquer que toutes ces propositions et bien des choses qu'on dit ici pourraient être conçues et énoncées plus généralement selon le style des géomètres. Par exemple, on pourrait dire en général que* les forces des corps sont en raison composée de la simple de leurs masses et de la doublée de leur vitesse, *au lieu que les quantités de mouvement sont en raison composée de la simple des masses aussi bien que des vitesses. Mais on s'est contenté de s'énoncer en certains nombres pour parler plus intelligiblement à l'égard de ceux qui sont moins accoutumés aux phrases des géomètres.*

État de l'autographe:
1.... *ce qu'il fallait* prouver.

Proof

Let A be the first weight, and B the second, and assume that A can rise to a certain height, for example, one foot. B will be able to rise to a height of 16 feet (Proposition 1); therefore, B has the force to raise one pound, namely, its own weight, to a height of 16 feet. And consequently (Axiom 2) it has the force to raise 16 pounds to one foot whereas A has only the force to raise 4 pounds, that is to say, its own weight, to 1 foot (by hypothesis). Now the force to raise 16 pounds to one foot is four times the force to raise 4 pounds to one foot (by common sense), therefore the force of B is four times the force of A. *Q.E.D.*

Proposition 7

A body of 4 pounds weight and one degree of velocity has the same force as a body of one pound weight and two degrees of velocity. And consequently if all the force of the former should be transferred to a body of one pound, it will receive only two degrees of velocity.

Proof

Let A be the first, and B the second. If A can rise to 1 foot, or raise 4 pounds, that is to say, its weight to 1 foot, B will be able to rise to 4 feet (Proposition 1) or raise its weight, which is 1 pound, to 4 feet. Therefore (Axiom 2), the force A is equal to that of B.[1]

Scholium

These two propositions can also be proved independently of Axiom 2, by Axiom 1 alone in conjunction with Definition 1, and using a procedure similar to that of Proposition 2 in order to reduce that [procedure] which would express the contrary to mechanical perpetual motion. Furthermore, we have pointed out in Axiom 2 that it can be proved by Axiom 1, that is to say, by reducing the contrary to perpetual motion, or *ad absurdum*. It is also worth noting that all these propositions and many things that are said here could be made known and stated more generally according to the manner of geometers. For example, we could say in general that *the forces of bodies are in a proportion compounded of the simple proportion of their mass and of the duplicate proportion of their weight*, whereas the quantities of motion are in a proportion compounded of the simple proportion of the masses as well as of the velocities. But we have been content to make statements in certain numbers in order to speak more intelligibly for those who are less accustomed to the expressions of geometers.

Original manuscript has:
1. ... *Q.E.D.*

Proposition 8

Lorsque les forces sont égales les quantités de mouvement ne sont pas toujours égales et vice versa.

Démonstration

Livres 4, vitesse 1 et livre 1, vitesse 2, sont d'une force égale (proposition 7), mais la quantité de mouvement de celui-là est double de la quantité de mouvement de celui-ci (définition 2). Vice-versa, livres 4, vitesse 1 et livre 1, vitesse 4, sont d'une quantité de mouvement égale (définition 2) mais la force de celui-là est seulement le quart de la force de celui-ci (proposition 6), et il en est de même en d'autres nombres.

Proposition 9

La même quantité de mouvement ne se conserve pas toujours.

Démonstration

Supposé que la même quantité de mouvement se conserve toujours, on peut obtenir le mouvement perpétuel mécanique (proposition 5), or ce mouvement est impossible (axiome 1), donc la même quantité de mouvement ne se conserve pas toujours.

Scholie

On le peut conclure encore de la proposition 8 et en effet, quand on s'opiniâtrerait à nier la deuxième demande, ou le deuxième postulatum sur lequel est fondée la proposition 5, c'est-à-dire quand on voudrait nier que toute la force d'un grand corps peut être transférée sur un petit corps (ce qui doit pourtant arriver souvent dans la nature) on n'éviterait pas pour cela la force de nos raisons. Car puisqu'on voit bien qu'ordinairement la quantité de mouvement est différente lorsque la force est la même, et vice versa (proposition 8) et que toujours la même force se doit conserver, afin qu'il n'y ait jamais un échange entre deux états dont l'un substitué à l'autre pourrait donner un mouvement perpétuel, il s'ensuit que le plus souvent la quantité de mouvement ne se conserve pas la même, soit qu'on transfère toute la force d'un corps sur un autre qui lui est inégal, ou qu'on en transfère une partie et en retienne l'autre. Ce que les géomètres prévoient d'abord à cause de la différence qu'il y a entre la raison simple et la raison doublée. Voyez le scholie de la proposition 7. En voici une preuve analytique générale pour leur satisfaction. Supposons que deux corps A et B se rencontrent avec les vitesses C et V et qu'après le choc ils aient les vitesses c et v. Donc si les quantités de mouvement se conservent, il faut qu'il y ait

Proposition 8

When the forces are equal, the quantities of motion are not always equal, and *vice versa.*

Proof

Pounds 4, velocity 1 and pounds 1, velocity 2, are of equal force (Proposition 7), but the quantity of motion of the former is double of the quantity of motion of the latter (Definition 2). *Vice versa*, pounds 4, velocity 1 and pounds 1, velocity 4, are of the same quantity of motion (Definition 2), but the force of the former is only one quarter of that of the latter (Proposition 6), and the case is the same for other numbers.

Proposition 9

The same quantity of motion is not always conserved.

Proof

It is assumed that the same quantity of motion is always conserved, then we can obtain mechanical perpetual motion (Proposition 5); now, such motion is impossible (Axiom 1), therefore, the same quantity of motion is not always conserved.

Scholium

We can deduce it moreover from Proposition 8, and indeed, if we should persist in denying the second postulate on which Proposition 5 is based, that is to say, if we should deny that all the force of a large body can be transferred to a small body (as must nevertheless often happen in nature), we should not be able to escape the force of our arguments on that account. For, as it is clear that the quantity of motion is usually different when the force is the same, and *vice versa* (Proposition 8), and that the same force must be conserved always in order that there should never be an exchange between two states, one of which when substituted for the other would be able to give perpetual motion, it follows that the quantity of motion is not conserved unchanged in most cases, whether all the force be transferred from one body to another which is not equal to it, or whether a part of it be transferred and part be retained. As for what the geometers foresee at first on account of the difference that exists between simple proportion and duplicate proportion, see the scholium to Proposition 7. Here is a general analytical proof for their satisfaction. Let two bodies A and B collide with velocities C and V, and let the velocities after collision be c and v. Then, if

AC + *BV égale à Ac* + *Bv, mais si les forces se conservent, il faut qu'il y ait ACC* + *BVV égal à Acc* + *Bvv, mais il est manifeste que ces deux équations ne se sauraient trouver véritables toutes deux qu'en certaine rencontre particulière, qu'il y a même moyen de déterminer. Et voici la détermination pour trancher court.* Deux corps se choquant directement ne sauraient conserver ensemble après le choc tant la somme de leurs forces, que la somme de leurs quantités de mouvement qu'ils avaient avant le choc, que lorsque la différence des vitesses avant le choc est égale à la différence réciproque des vitesses après le choc. *Toutes les fois que les corps vont d'un même côté, tant avant qu'après le choc, cela arrive.*

Remarques

La considération de l'équilibre a contribué beaucoup à confirmer les gens dans cette opinion, qui paraissait vraisemblable d'elle-même, que la force et la quantité de mouvement reviennent à la même chose, et que les forces sont égales lorsque les quantités de mouvement sont égales, c'est-à-dire lorsque les vitesses sont réciproquement comme les poids, et qu'ainsi la force de livres 4, vitesse 1, est égale à celle de livre 1, vitesse 4. Car on voit qu'il se fait équilibre toutes les fois que les poids sont disposés en sorte que l'un ne peut descendre sans que l'autre monte avec des vitesses réciproques aux poids. Mais il faut savoir que cela y réussit comme par accident, car il arrive alors qu'encore les hauteurs de la montée ou de la descente sont réciproques aux poids. Or c'est une règle générale qui se déduit par des raisons que nous venons de proposer, que les forces sont en raison composée des poids et des hauteurs auxquels ces poids se peuvent élever en vertu de leurs forces. *Et il est à propos de considérer que l'équilibre consiste dans un simple effort* (conatus) *avant le mouvement, et c'est ce que j'appelle la* force morte *qui a la même raison à l'égard de la* force vive (*qui est dans le mouvement même*) *que le point à la ligne. Or au commencement de la descente lorsque le movement est infiniment petit, les vitesses ou plutôt les éléments des vitesses sont comme les descentes, au lieu qu'après l'élévation, lorsque la force est devenue vive, les descentes sont comme les carrés des vitesses. Il y a encore une chose qui mérite d'être-observée. C'est qu'un globe de 4 livres de poids et d'un degré de vitesse et un autre globe d'une livre de poids et de 4 degrés de vitesse quand ils se rencontrent directement s'empêchent mutuellement d'avancer, comme dans l'équilibre, et qu'ainsi, quant à l'effet d'empêcher l'avancement, ils ont une même force respective. Mais cependant leurs forces absolues sont bien inégales, puisque l'un peut produire 4 fois autant d'effet que l'autre. Voyez proposition 6. Or il s'agit ici de la force vive et absolue. Ces variétés paradoxes ont contribué beaucoup à embrouiller la matière, d'autant qu'on n'a pas eu des idées bien distinctes de la force et de ses différences. Mais*

the quantities of motion be conserved, $AC + BV$ must equal $Ac + Bv$; but if the forces be conserved, $ACC + BVV$ must equal $Acc + Bvv$; but it is manifest that these two equations could not both be true except in the case of a particular collision, which can be determined. Here is the method in brief. *Two bodies colliding directly would not be able to conserve together after collision such sum of their forces as the sum of the quantities of motion that they possessed before collision, except when the difference of the velocities before collision is equal to the reciprocal difference of the velocities after collision.* That happens whenever the bodies come from the same side, as well before as after collision.

Remarks

Consideration of equilibrium has contributed much to confirm individuals in that opinion, which seemed probable in itself, that force and quantity of motion come to the same thing, and that forces are equal when the quantities of motion are equal, that is to say, when the velocities are reciprocally proportional to the weights; thus, the force of pounds 4, velocity 1, is equal to that of pounds 1, velocity 4. For, it is clear that equilibrium is established whenever the weights are so disposed that one cannot fall down without the other rising with velocities reciprocally proportional to the weights. But it must be realized that that happens as though by accident, for it happens moreover when the distances of rise or of fall are reciprocally proportional to the weights. Now, it is a general rule, which is deduced by the arguments that we have just put forward, *that forces are in a proportion which is compounded of the weights and of the heights to which those weights can be raised in virtue of their forces.* And it is reasonable to consider that equilibrium consists in a simple effort (*conatus*) before motion, and that is what I call *potential force* which has the same relationship with respect to *kinetic force* (which is the motion itself) as does a point to a line. Now, at the beginning of the fall when the motion is infinitesimal, the velocities, or rather the elements of the velocities are proportional to the distances traversed, whereas, after being raised to a height, when the force has become kinetic, the distances traversed in falling are proportional to the squares of the velocities. There is one thing further that is worth noting. When a sphere of 4 pounds weight and one degree of velocity collides directly with another sphere of one pound weight and 4 degrees of velocity they mutually prevent each other from progressing, as in equilibrium, and thus, as regards the effect which prevents progression, they have the same relative force. But, however, their absolute forces are definitely unequal, seeing that the one can produce

127

j'espère que dans nos Dynamiques on trouvera ces choses éclairées à fond.

Si quelqu'un veut donner un autre sens à la force, comme en effet on est assez accoutumé à la confondre avec la quantité de mouvement, je ne veux pas discuter sur les mots et je laisse aux autres la liberté que je prends d'expliquer les termes. C'est assez qu'on m'accorde ce qu'il y a de réel dans mon sentiment, savoir que ce que j'appelle la force *se conserve, et non pas ce que d'autres ont appelé de ce nom. Puisque autrement la nature n'observerait pas la loi de l'égalité entre l'effet et la cause et ferait un échange entre deux états, dont l'un substitué à l'autre pourrait donner le mouvement perpétuel mécanique, c'est-à-dire un effet plus grand que la cause.*

On pourrait aussi donner une autre interprétation à la quantité de mouvement selon laquelle cette quantité se conserverait, mais ce n'est pas celle que les Philosophes ont entendue. Par exemple les corps A et B allant chacun avec sa vitesse, la quantité totale du mouvement est la somme de leurs quantités de mouvement particulières, comme la force totale est la somme de leurs forces particulières; et c'est ainsi que Descartes et ses sectateurs ont entendu la quantité de mouvement, et pour en être assuré on n'a qu'à voir les règles du mouvement que lui ou d'autres, qui ont suivi son principe, ont données. Mais si l'on voulait entendre par la quantité de mouvement, non pas le mouvement absolument pris (où l'on n'a point égard de quel côté il va) mais l'avancement vers un certain côté, alors l'avancement total (ou le mouvement respectif) sera la somme des quantités de mouvement particulières, quand les deux corps vont d'un même côté. Mais lorsqu'ils vont l'un contre l'autre, ce sera la différence de leurs quantités de mouvement particulières. Et on trouvera que la *même quantité d'avancement se conserve. Mais c'est ce qu'il ne faut pas confondre avec la quantité de mouvement prise dans le sens ordinaire. La raison de cette maxime de l'avancement paraît en quelque façon et il est raisonnable que rien ne survenant du dehors, le tout (composé des corps en mouvement) ne s'empêche pas lui-même d'avancer autant qu'il faisait. Mais j'en ai donné ailleurs une démonstration exacte.*

Il est encore à propos de remarquer que la force se peut estimer sans faire entrer le temps dans la considération. Car une force donnée peut produire un certain effet limité qu'elle ne surpassera jamais quelque temps qu'on lui accorde. Et soit qu'un ressort se débande tout d'un coup ou peu à peu, il n'élèvera pas plus de poids à la même hauteur, ni le même poids plus haut. Et un poids qui monte en vertu de sa vitesse n'arrivera pas plus haut, soit qu'il monte perpendiculairement, ou qu'il monte obliquement dans un plan incliné ou bien dans une ligne courbe. Il est vrai que la montée oblique demande plus de temps pour arriver à la même hauteur, mais elle fait aussi plus de chemin et plus de détours. De sorte que pour estimer la force par le temps il faut aussi considérer tous les chemins et tous les détours. Mais on est dégagé de tout cela quand on considère le seul effet qui se peut produire

4 times as much effect as the other. See Proposition 6. Now, we are concerned here with kinetic and absolute force. These paradoxical differences have contributed much to confuse the matter, especially as there have been no very distinct ideas about force and its variations. But I hope that these things will be found to be thoroughly explained in our Dynamics.

If anyone wishes to give another meaning to force, for indeed we are not unaccustomed to its being confused with quantity of motion, I do not wish to argue about words and I leave to others the freedom that I take in explaining terms. It is enough that I be conceded what is really in my mind, namely, that what I call *force* is conserved, and not what others have called by that name. Because, otherwise, nature would not obey the law of equality between cause and effect and would make an exchange between the two states, one of which when substituted for the other would be able to give mechanical perpetual motion, namely, an effect greater than the cause.

It would be possible also to give another explanation of quantity of motion, according to which that quantity would be conserved, but it is not what is meant by Philosophers. For example, in the case of bodies *A* and *B*, each moving with its own velocity, the total quantity of motion is the sum of their individual quantities of motion; and that is what Descartes and his followers have understood by quantity of motion, and to be convinced of the fact, it is only necessary to look at the rules of motion which he or others, who have adopted his principle, have given. But if we wanted quantity of motion to mean, not motion taken absolutely (where no attention is paid to the direction it takes), but progression in a certain direction then the total progression (or relative motion) will be the sum of the individual quantities of motion, when the two bodies come from the same direction. But when they come from opposite directions, it will be the difference of their individual motions. And it will be found that the *same* quantity of progression is conserved. But that must not be confused with the quantity of motion taken in the usual sense. The ground for this maxim of progression is rather apparent, and it is reasonable, when nothing happens from without, that everything (composed of moving bodies) should not be hindered from progressing by itself as much as before. But I have given an exact proof of that elsewhere.

It is furthermore apposite to remark that force can be evaluated without taking time into consideration. For a given force can produce a certain limited effect which it can never exceed whatever time be allowed to it. Let a spring uncoil itself suddenly or gradually, it will not raise a greater weight to the same height, nor the same weight to a greater height. And a weight which rises in virtue of its velocity will not attain a

après tous ces détours. C'est ainsi, par exemple, qu'on prévoit d'abord, sans avoir presque besoin de démonstration ou de raisonnement, que le jet d'eau, libre de tous empêchements accidentels, doit jaillir précisément à la hauteur de l'eau, ou à la surface supérieure. Car c'est afin que l'eau puisse précisément retourner par l'ouverture d'en haut dans le vase d'où elle sort par la lumière d'en bas, et continuer toujours le même jeu par un mouvement perpétuel physique, tout comme un pendule parfaitement libre doit remonter précisément à la hauteur d'où il est descendu, autrement l'effet entier ne serait pas égal à sa cause totale. Mais comme il est impossible d'exclure tous les empêchements accidentels, ce jeu cesse bientôt dans la pratique, autrement ce serait le mouvement perpétuel mécanique. Cependant cette considération nous donne une voie abrégée pour estimer les effets par les forces, ou les forces par les effets, et pour connaître les véritables lois de la nature.

Il y a déjà eu quelques habiles hommes de ce temps qui ont trouvé par des expériences ou raisons particulières que la quantité de mouvement ne saurait se conserver toujours. Mais comme on était prévenu de l'opinion que la quantité de mouvement est la même chose que la force, ou qu'au moins les forces sont comme les quantités de mouvement en raison composé des masses et des vitesses, et qu'ainsi l'accroissement de la vitesse récompense précisément le décroissement de la masse on avait de la peine à se rendre à leurs raisons qu'on soupçonnait d'être fausses. Car on ne pouvait comprendre comment une partie de la force pouvait être perdue sans être employée à rien, ou gagnée sans venir de rien. On considérait la masse comme de l'eau et la vitesse comme du sel qu'on faisait dissoudre dans cette eau, et l'on concevait bien le sel plus étendu dans plus d'eau, ou plus resserré dans moins d'eau, et même tiré d'une eau et transféré dans une autre. Mais j'ai déjà fait voir comment en cela on a péché contre la métaphysique réelle, et contre la science d'estimer les choses en général.

Maintenant que la véritable notion de la force est établie, et que la source tant de l'erreur que de la vérité, est découverte, on sera plus disposé à se désabuser. Tout cela est d'autant plus raisonnable que le mouvement est une chose passagère qui n'existe jamais à la rigueur puisque ses parties ne sont jamais ensemble. Mais c'est la force (qui est la cause du mouvement) qui existe véritablement, ainsi outre hors de la masse, de la figure et de leur changement (qui est le mouvement) il y a quelque autre chose dans la nature corporelle: savoir la force. Il ne faut donc pas s'étonner si la nature (c'est-à-dire la sagesse souveraine) établit ses lois sur ce qui est le plus réel.

greater height, whether it rise perpendicularly, or whether it rise obliquely on an inclined plane or even on a curve. It is true, that the inclined rise requires more time to arrive at the same height, but it takes also a longer path and makes more deviations. So, in order to evaluate force by time, it is necessary to consider in addition all paths and all deviations. But all that is obviated when we consider only the effect that is produced after all the deviations. Thus, for example, we foresee straight away, without having recourse to proof or argument, that a fountain, free from all accidental hindrances, must spout exactly to the height of the water, or to the upper surface. The reason being that the water must return as it happens through the opening at the top to the vessel from which it came through the opening at the bottom and must always continue the same play by a physical perpetual motion, exactly in the same way that a perfectly free pendulum must rise to exactly the height from which it fell, otherwise the whole effect would not be equal to its total cause. But as it is impossible to exclude all accidental hindrances, that movement soon ceases in practice, otherwise that would be mechanical perpetual motion. Nevertheless, that consideration provides us with a short way of evaluating effects by forces, or of forces by effects, and to ascertain the true laws of nature.

Already, some able men of our time have found by experiment or by special arguments that the quantity of motion could not always be conserved. But having been previously informed of the opinion that quantity of motion is the same thing as force, or at least that forces are proportional to quantities of motion compounded of mass and velocity, and consequently that an increase in velocity is exactly compensated by a decrease in mass, they had difficulty in accounting for their arguments which they suspected were false. For, they could not understand how one part of the force could be lost without doing something, or gained without coming from somewhere. Mass was looked upon in the same way as water, and velocity in the same way as salt which was dissolved in water; and they regarded salt as being more extended in more water, or more confined in less water, and even withdrawn from one water and transferred to another. But I have already pointed out how that conflicts with real metaphysics, and with the appraisal of things in general.

Now that the true concept of force is established, and that the source of error as well as of truth is revealed, we shall be more disposed to disillusion ourselves. That is all the more reasonable seeing that motion is a transient thing which never exists strictly speaking, seeing that its parts are never all together. But it is force (which is the cause of motion) that truly exists; so, in addition, apart from mass, shape and change (which is motion), there is something else in corporeal nature: namely, *force*. Consequently, we must not be surprised if nature (that is to say, the Divine Wisdom), established its laws on that which is most real.

131

Appendix II

GENERAL RULE FOR THE COMPOUNDING OF MOTIONS

The differences between the two versions of the accounts written by Leibniz are shown in the relevant portions reproduced below. See beginning of chapter III.

Article from *Journal des Sçavans,* 7 September 1693, p. 417

Si les droites AB, AC, AD, AE, etc., représentent les diverses tendences ou les mouvements particuliers *d'un* mobile A

....
....
....................B, C, D, E, etc.,
....
....
....
....
....
....
.....

d'A jusqu'à B, en cas qu'il eut été poussé par le seul mouvement AB que je suppose toujours uniforme ici) et encore de même s'il était parvenu dans une seconde jusqu'à C, ou D, ou E, etc., en cas qu'il eust été poussé par un de ces mouvements tout seul; maintenant que ce mobile...........
....
....
....
de tous les points de tendence B, C, D, E, etc., d'autant plus loin qu'il y a plus de tendences, de sorte qu'il parviendra dans une seconde jusqu'à M si AM est à AG comme le nombre des tendences est à l'unité. *Ainsi il arrivera au mobile la même chose qui arriverait à son centre de gravité, si ce mobile se partageait également entre ces mouvements pour satisfaire parfaitement à tous ensemble.* Car le mobile étant partagé également entre 4 tendences, il ne peut échoir à chacune qu'une quatrième partie du mobile qui devra aller quatre

Copy made by Des Billettes 1692

Si les droites AB, AC, AD, etc., représentent les mouvements particuliers du mobile A

qui doivent composer un mouvement total et si G est le centre de gravité de *tous les points de tendence* B, C, D, etc.

Enfin si AG est prolongé jusqu'à M en sorte qu'AM soit à AG comme le nombre des mouvements composants est à l'unité, le mouvement composé sera AM. C'est-à-dire, pour parler plus familièrement, si le mobile A serait parvenu dans une seconde de temps
d'A à B, ou à C ou à D, etc., supposé qu'un de ces mouvements eut été seul,

maintenant que le mobile est poussé en même temps par tous ces mouvements ensemble, *ne pouvant pas aller en même temps de plusieurs côtés il ira vers le centre de gravité* de tous les points de tendence,

et avec une telle vitesse, qu'il parviendra dans une seconde jusqu'à M *de sorte qu'il arrivera au mobile la même chose qui arriverait à son centre de gravité si le mobile se partageait également entre tous ces mouvements, et allait selon chacun d'autant plus vite qu'il serait devenu plus petit par le partage.*

fois plus loin, pour avoir autant de progrès que si le mobile tout entier avait satisfait à chaque tendence. Mais ainsi le centre de gravité de toutes ces parties irait aussi quatre fois plus loin. Maintenant le partage n'ayant point de lieu, le tout ira comme le centre des partages, pour satisfaire à chaque tendence en particulier, autant qu'il est possible dans le partage. Et il en provient autant que si on avait fait les partages, et réuni les parties au centre, après avoir satisfait aux mouvements particuliers.

Cette explication peut tenir lieu de démonstration. Mais ceux qui en demandent une à façon ordinaire, le trouveront aisément en poursuivant ce qui suit.

Si l'on mène par A deux droites qui soient dans un même plan avec tous les mouvements et qui fassent un angle droit en A, on pourra..............
....
....
....
....
....
....
....
....
....
....
....
....
....
....
....
....
....
divisée par leur nombre: observant que ce qui est en sens contraire est une quantité négative, dont l'addition est une soustraction en effet. Or puisqu'il faut multiplier par le nombre des tendences la distance du centre de gravité des points de tendence, pris tant sur l'un que sur l'autre côté de l'angle droit, pour déterminer le mouvement composé sur chacun des côtés, il s'ensuit que le mouvement total composé des mouvements de ces deux côtés, se déterminera de même. Ainsi la composition de plusieurs mouvements faisans angle ensemble dans un même plan, se réduit

Cette explication peut tenir lieu de démonstration. Car ainsi il se fait autant de progrès qu'auparavant, mais ceux qui en demandent une démonstration à la façon ordinaire la trouveront aisément en poursuivant ce qui suit.

Si l'on mène par A deux droites qui fassent un angle droit en A, on pourra résoudre chacun de tous ces mouvements particuliers en deux pris sur les côtés de cet angle droit. Ainsi la composition de tous ces mouvements sur un des côtés sera le mouvement moyen arithmétique multiplié par le nombre des mouvements. C'est-à-dire, pour avoir la distance entre A et le point de tendence de ce mouvement sur ce côté il faudra multiplier la distance du centre de gravité de tous les points de tendence sur le même côté par le nombre des tendences. Car on sait que la distance entre A et le centre de gravité des points pris sur une même droite avec A, est la moyenne arithmétique des distances entre A et ces points, de quelque nombre qu'ils puissent être. J'appelle une *grandeur moyenne arithmétique* entre plusieurs grandeurs, celle qui se fait par leur somme divisée par leur nombre.

Ainsi le *mouvement composé* sur chacun des côtés étant *celui du centre de gravité des partages, ou replications*

133

à la composition de plusieurs mouvements dans une même droite et de deux mouvements faisans angle droit.

Que si les mouvements donnés ne sont pas dans le même plan, il faut se servir de trois droites faisans angle entre elles.

Il est bon de remarquer que dans cette composition des mouvements, il se conserve toujours la même quantité de progression, et non pas toujours la même quantité de mouvement. Par exemple si deux tendences sont dans une même droite, mais en sens contraire, le mobile va du côté du plus forte avec la différence des vitesses, et non point du tout avec leur somme, comme il arriverait si les tendences le portaient d'un même côté.

Et si les deux tendences contraires étaient égales, il n'y aurait point de mouvement. Cependant cela suffit, pour ainsi dire, *in abstracto* lorsqu'on suppose déjà ces tendences dans le mobile, mais *in concreto* en considérant les causes qui les y doivent produire, on trouvera qu'il ne se conserve pas seulement en tout la même quantité du progrès, mais aussi la mème quantité de la force absolue et entière, qui est encore différente de la quantité du mouvement.

On donnera une autre fois *deux Consectaires* fort généraux et fort importants qui se tirent de cette règle.

du mobile multiplié par leur nombre, le même arrivera du mouvement diagonal composé de celui des deux côtés qui est le mouvement composé total.

De cette règle se tirent *deux Consectaires* importants, savoir des constructions aisées de deux problèmes que voici:

Journal des Sçavans
14 September 1693, p. 423

Problème A. — Mener la tangente d'une ligne courbe qui se décrit par des *filets tendus.*

Du point A de la courbe . . .

. . . .
. . . .
. . . .
. . . .
. . . .
. . . .
. . . .
. . . .
. . . .
. . . .
. . . .

... d'autant plus pesant. On peut appliquer cette construction non seulement aux coniques ordinaires, aux ovales de M. Descartes, aux coévolu-

Problème 1. — Mener la tangente d'une ligne courbe qui se décrit par des filets tendus à la façon des coniques ou des ovales de M. Descartes.

Du point A de la courbe soit décrit un cercle quelconque coupant les filets aux points B, C, D, etc., soit trouvé le centre de gravité de ces points, G, et AG sera perpendiculaire à la courbe, ou bien, une droite menée par A normale à AG sera la tangente que l'on cherche. Lorsque le filet est double ou triple il y faut considérer deux ou trois points dans un seul endroit, à peu près comme si un de ces points tenant lieu de plusieurs était d'autant plus pesant.

tions de M. de Tchirnhaus, mais encore à une infinité d'autrés lignes. En voici la raison, qui a servi de principe d'invention............................
....
....
... qu'il y a de filets: car il les tire également et comme il les tire, il en est tiré...............................
....
....
... filets (par la nouvelle règle des compositions du mouvement que l'auteur vient de publier dans le journal précédent). Et ces points.................
....

M. de Tchirnhaus, dans son livre intitulé *Medicina mentis* ayant cherché le premier ce problème a donné occasion à M. de Leibniz d'y arriver; ce qu'il a fait en prenant une voie qui a cet avantage que l'esprit y fait tout sans calcul et sans diagrammes.

M. Facio y est aussi arrivé de son chef par une très belle voie et l'a publié le premier. Enfin M. le marquis de l'Hôpital a donné sur ce sujet l'énonciation la plus générale qu'on puisse souhaiter, fondée sur la nouvelle méthode du *Calcul des différences*.

Problème Z. — ....
....
....
.................J'appelle sollicitations les efforts infiniment petits ou *conatus*, par lesquels le mobile est sollicité ou invité, pour ainsi dire, au mouvement; comme est par exemple l'action de la pesanteur ou de la tendence centrifuge, dont il en faut une infinité pour composer un mouvement ordinaire.
....
....
....
....
....
....
....

Le problème qu'on vient de résoudre est d'importance en Physique; car la nature ne produit jamais aucune action que par une multitude véritablement infinie des causes concourantes.

Voici la raison de tout ceci qui servit de principe d'invention. On doit considérer que le stile qui tend les filets pourrait être conçu comme ayant autant de directions égales en vitesse entre elles, qu'il y a de filets. Car comme il les tire, il en est tiré. Ainsi la direction composée qui doit être dans la perpendiculaire à la courbe, passe par le centre de gravité d'autant de points qu'il y a de filets. Et ces points à cause de l'égalité des tendences sont également distants du stile, et tombent ainsi dans les intersections du cercle avec les filets.

M. de Tschirnhaus fut le premier qui tâchât de trouver quelque règle pour les tangentes des courbes décrites avec des filets et cela donna occasion à M. d. L. de la chercher aussi, ce qu'il fit avec succès par la voie que nous venons de dire, mais comme il ne se hâtait guère de publier ses pensées, M. Facio dont nous avons de très beaux essais en mathématiques trouva à peu près la même chose là-dessus, et M. Huygens y contribua, mais c'est par une voie bien différente de celle-ci.

Problème 2. — Un même mobile étant poussé en même temps par un nombre infini de sollicitations, trouver son mouvement. J'appelle *sollicitations* tous les efforts infiniment petits,

comme est celui de la pesanteur, ou encore celui de la force centrifuge, et il en faut une infinité pour composer un mouvement ordinaire.

Cherchez le centre de gravité du lieu de tous les points de tendence de toutes ces sollicitations, et la direction composée passera par ce centre, mais les vitesses produites seront proportionnelles aux grandeurs des lieux. Les lieux peuvent être des lignes, des surfaces, ou même des solides.

Il est bon aussi de considérer qu'ici où il ne s'agit que de la quantité de la progression, les vitesses et les grandeurs se récompensent. Et par ce moyen on pourrait diversifier les mouvements car il se conserve toujours la même quantité de progression, mais on a démontré ailleurs que lorsqu'il s'agit de la force absolue ils ne se compensent point, car il ne se conserve pas toujours la même quantité de mouvement total.

Name Index

Subject Index

Items in italic type relate to concepts particularly adopted by Leibniz

138

Bibliography

Bernoulli, Johann: *Der Briefwechsel von J. Bernoulli.* O. Spiess, Bâle, 1955.

Catelan: See *Nouvelles de la République des Lettres,* 1687.

Costabel, Pierre: *Le Mouvement,* in *Encyclopédie Clartés,* vol. XVI, fasc. 16150–16165. Paris, 1956.

———— 'La démonstration cartésienne relative au centre d'équilibre de la balance', in *Archives Internationales d'Histoire des Sciences,* 1956, IX, 133 *et seq.*

———— 'La controverse Descartes-Roberval au sujet du centre d'oscillation', in *Revue des Sciences Humaines,* 1951, pp. 74–86.

———— *Centre de Gravité et Équivalence dynamique* (*Conférences du Palais de la Découverte,* December 1954, Series D, No. 34).

———— 'Deux Inédits de la Correspondance indirecte Leibniz-Reyneau, in *Revue d'Histoire des Sciences',* 1949, II, 311–332.

———— 'La septième Règle du Choc Élastique de Christian Huygens', in *Revue d'Histoire des Sciences,* 1957, X, fasc. 2, p. 120.

Descartes, René: *Œuvres de Descartes,* ed. Adam Tannery, Paris, Vrin, 1956.

Fatio de Duillier, Nicolas: See *Bibliothèque Universelle et Historique,* 1687–1689.

Foucher de Careil, *Comte* Alexandre Louis: See Leibniz.

Gerhardt, Carl Emmanuel: See Leibniz.

Guéroult, Martial: *Dynamique et Métaphysique Leibniziennes,* Paris, Les Belles Lettres, 1934.

Hannequin, Arthur Édouard: *La première philosophie de Leibniz.* (*Études d'histoire des Sciences et d'histoire de la philosophie,* vol. II. Paris, Alcan, 1908).

Hobbes, Thomas: *Philosophia Prima Opera Philosophica,* London, Bohn, 1839, Pars II.

Huygens, Christiaan: *Œuvres complètes de Christiaan Huygens,* La Haye, Société hollandaise des Sciences, 1888–1950.

Lamy, Bernard: *Traité de Mécanique,* 2nd ed. Paris, 1687.

Leibniz, G. W.: *Œuvres de Leibniz . . .,* ed. A. Foucher de Careil. Paris, Firmin-Didot, 1859–1875.

BIBLIOGRAPHY

——— *Lettres et Opuscules inédits de Leibniz*, ed. A. Foucher de Careil, Paris, Lagrange, 1854.

——— *Der Briefwechsel von G. W. Leibniz mit Mathematikern*, ed. Gerhardt, vol. I, Berlin, 1899.

——— *Die philosophischen Schriften von G. W. Leibniz*, ed. Gerhardt, Berlin, 1875–1890.

——— *Leibnizens mathematische Schriften...*, ed. Gerhardt, Berlin, 1849–1863.

——— *Opera omnia*, ed. Dutens, Geneva, 1678.

L'Hôpital: See Bernoulli.

Mouy, Paul: *Le développement de la Physique Cartésienne (1646–1712)*, Paris, Vrin, 1934.

Pardies, Ignace Gaston: *La Statique ou la Science des forces mouvantes*, Paris, 1673.

Pascal, Blaise: *Lettres de A. Dettonville contenant quelques-unes de ses inventions de géométrie*, Paris, G. Desprez, 1659.

Tschirnhaus, Ehrenfried Walther von. *Medicina mentis seu tentamen genuinae logicae...*, Amsterdam, 1687.

Varignon, Pierre: *Projet d'une nouvelle Mécanique*, Paris, 1687.

——— *Nouvelle Mécanique*, posthumous ed. Paris, 1725.

Robinet, André: *Malebranche et Leibniz, relations personelles*, Paris, Vrin, 1955.

PERIODICALS CONSULTED

Journal des Sçavans, 1682, 1692, 1693.
Journal des Savants, 1844.
Nouvelles de la République des Lettres, 1687, 1705.
Bibliothèque Universelle et Historique, 1687–1689.

UNPUBLISHED MANUSCRIPTS

Archives of the *Académie des Sciences*, Paris.
Bibliothèque de l'Institut de France.
Landesbibliothek, Hanover

BR Correspondence	BR Toinard
HS Autographs	BR Larroque
LBH MS Manuscripts	BR Malebranche
	BR Foucher
	HS Theology

Printed in Great Britain by
William Clowes & Sons, Limited
London, Beccles and Colchester